River Basin Management - Challenges and Coping Approaches

Edited by Chao Guo

Published in London, United Kingdom

River Basin Management - Challenges and Coping Approaches
http://dx.doi.org/10.5772/10.5772/intechopen.1003543
Edited by Chao Guo

Contributors
Belen Marti-Cardona, Chao Guo, Chau Ha Pham, Chizya Mvula, Fauziah Ahmad, Frank Karlsen, Giriraj Amarnath, Lars Eric Roseng, Leila Tajedin, Li Zhou, Mandana Abedini, Md Azlin Md Said, Moses N. Chisola, Neda Khademi Shiraz, Nivedhitha Jothinarayanan, Salar Khani, Sisay E. Debele, Yakob Umer

First published in London, United Kingdom, 2025 by IntechOpen
IntechOpen is the global imprint of INTECHOPEN LIMITED, registered in England and Wales, registration number: 11086078, 167-169 Great Portland Street, London, W1W 5PF, United Kingdom

For EU product safety concerns: IN TECH d.o.o., Prolaz Marije Krucifikse Kozulić 3, 51000 Rijeka, Croatia, info@intechopen.com or visit our website at intechopen.com.

British Library Cataloguing-in-Publication Data
A catalogue record for this book is available from the British Library

River Basin Management - Challenges and Coping Approaches
Edited by Chao Guo
p. cm.

This title is part of the Earth Sciences Book Series, Volume 4
Series Editor: Maurizio Lazzari

Print ISBN 978-0-85466-260-9
Online ISBN 978-0-85466-259-3
eBook (PDF) ISBN 978-0-85466-261-6
ISSN 3049-8848

IntechOpen Book Series
Earth Sciences
Volume 4

Aims and Scope of the Series

The world of Earth Sciences, considering the interactions within the geosphere and between the geosphere–biosphere, is a place in which a large number of scientists find and have found over time their own relationship or sector of application precisely because it is absolutely transversal to many disciplines and subdisciplines, which do not necessarily fall within the Geosciences. The objective of this book series is to welcome original scientific contributions both in consolidated contexts and in new frontiers of research, as well as review papers included in the various disciplines of Earth Sciences, but above all, those that show a modern and transversal vision of applications and impacts on the community in a particular historical context, which, following the COVID-19 pandemic, has shifted global attention to sectors that were previously more neglected. In particular, those of mining research and fossil and renewable energy sources, environmental geology and the sustainable use of natural resources and impacts on the built environment, land consumption, geoarchaeology, forensic geology, geotourism/geoheritage, georisks and climate and environmental changes, considered at different scales, up to new applications of geostatistical and geospatial analysis, GIS and artificial intelligence for the definition of forecasting models and scenarios in various sectors of basic and applied research.

Meet the Series Editor

Dr. Maurizio Lazzari has a Ph.D. in Earth Science and is a researcher at the Italian National Research Council, Institute of Cultural Heritage Sciences. Since 2001, Dr. Lazzari has been a Professor of Pedology at the University of Basilicata (Italy) and a geoarchaeologist at the University of Salento (Italy). His research activities are focused on natural and anthropic hazards and risk factors, aimed at safeguarding and conserving settlements and the historical–monumental heritage of the Mediterranean, with particular attention to landslide processes, susceptibility maps, monitoring, and modelling. Since 2004, he has been working as a scientific coordinator for several national research projects studying landslides and triggering factors, natural and anthropogenic risks, geological and geomorphological mapping, soil erosion, preservation of historical and archaeological sites, enhancement of degraded areas, geo-touristic use, and the protection of the landscapes. He is the author of about 150 scientific publications in national and international journals, monographs, book chapters, and conference proceedings concerning applied geology, geomorphology, dynamics of artificial reservoirs, soil erosion, landslides, geoarchaeology, hydrogeological instability, natural hazards, monitoring, cultural landscape, UNESCO Heritage, geoarchaeology, and geo-tourism.

Meet the Volume Editor

Dr. Chao Guo is a senior engineer and serves as a master's supervisor. His research interests are focused on fine-sediment dynamics, the effects of reservoirs on streamflow and sediment load, and river flood control and management. Dr. Guo has been the PI or the core participant for more than 20 national and provincial-level projects, including the Natural Science Foundation of China, the National Key Research & Development Program, etc. Dr. Guo has published more than 40 research articles, obtained 7 national patent authorizations, and participated in developing 2 technical standards by the Ministry of Water Resources of China. He has been awarded the Ministry of Water Resources' International Talent and the Young Top Talent in Hubei Province.

Contents

Preface

Rivers are among Earth's most important natural resources, playing a crucial role in supporting human societies and ecosystems. However, they are increasingly threatened by the combined impacts of climate change and anthropogenic activities. These pressures have led to extreme floods and droughts, accelerated soil erosion, declining water quality, and the degradation of aquatic ecosystems. Addressing these challenges and ensuring the sustainable development of rivers has become an urgent and critical task before us. This book addresses these issues by exploring the latest methods and achievements in river research and management, offering scientific insights and technical solutions to support effective river conservation and restoration. Besides the Introduction, the book focuses on three key sections:

1. Monitoring Technologies for Eco-Environmental Parameters

Monitoring water quality and aquatic ecosystems is essential for assessing river health and the effectiveness of pollution control measures. This section introduces advanced techniques, such as the integrated approach using wavelet analysis and machine learning, as well as environmental DNA (e-DNA) methods. These innovative technologies provide comprehensive data to support eco-environmental protection and management.

2. Improved Modeling Approaches

This section discusses enhancements to traditional watershed erosion models to improve the accuracy of erosion predictions, providing more reliable foundations for soil and water conservation measures. Additionally, it explores the application of big data and machine learning techniques to enhance the efficiency and precision of hydrodynamic models, offering more accurate decision-making support for river management.

3. Nature-Based Solutions

Traditional river engineering projects often negatively impact ecosystems. This section explores the principles and methodologies of nature-based river restoration, using the Kitwe region in Zambia as a case study. It highlights the effectiveness of these approaches in mitigating floods, reducing disaster risks, and enhancing ecosystem resilience.

This book is designed to serve as a comprehensive reference for researchers, engineers, and policymakers involved in river study, management, and conservation. By sharing the latest scientific insights and practical solutions, we aim to foster collaboration and inspire innovative approaches to river sustainability. Together, we can address our rivers' pressing challenges and work towards a more resilient and sustainable future.

Chao Guo
Changjiang River Scientific Research Institute of Changjiang Water Resources Commission,
Wuhan, China

Section 1

Introduction

Chapter 1

Introductory Chapter: Addressing Climate Change and Human Impacts for Sustainable River Basin Management

Chao Guo

1. The important role of rivers

Rivers are one of the most important natural resources on Earth, supporting indispensable functions for both humans and ecosystems [1]. Firstly, rivers are a primary source of freshwater, supplying essential water for agriculture, industry, and daily life. Secondly, rivers play a significant role in regulating climate and maintaining ecological balance, through cycling water, sediment, and nutrients. They also serve the functions of navigation, energy generation (such as hydropower), and recreational activities.

2. Climate change and human activities induced growing threats to rivers

Despite their importance, rivers globally are experiencing growing threats from climate change and human activities over the past two to three decades [2]. The global temperature rise is now expected to be between 1.5 and 5°C by the end of the century [3]. Global warming has accelerated the glacier melting and changed rainfall patterns, leading to frequent floods in some areas and more severe droughts in others. Moreover, rising water temperatures and the increase in extreme weather events have disrupted the balance of river ecosystems, endangering aquatic life [4]. It was reported that we are on the brink of an irreversible climate disaster [5]. Take the past year of 2024 as an example, in March–May, heavy rainfall in East Africa caused severe flooding that killed hundreds and affected more than 700,000 people. Then in April–May, extensive flooding in southern Brazil devastated 478 cities, killed 173 people. In October, rainstorm in eastern Spain led to catastrophic floods, resulting the death of over 220 people, marking the deadliest flood event in Europe since 1967. In addition to climate-induced effects, river basins are also under immense pressure from human activities [6]. Dam construction, wastewater discharge, land use changes, and excessive water extraction are major issues. It was estimated that the number of reservoirs to be >16 million globally, and dams alter natural river flows, affecting sediment and nutrients transport and fish migration, thus causing the loss of river connectivity and disrupting ecological balance [7]. Palmer et al. [8] highlighted that river basins affected by dams or extensive development are likely to undergo more significant changes in discharge and water stress compared to undisturbed rivers within the next 50 years. The discharge of industrial, agricultural, and domestic wastewater leads to river pollution, degrading water quality and threatening the survival of aquatic life. Urbanization and land use changes increase soil erosion, increasing sediment levels in rivers. There is also a risk that water extraction rates could rise in response to water scarcity, meaning even more rivers might dry up.

3. The need and methodology for river research

As climate change and human activities intensify, research on river substances transport and basin management has become critical for sustainable development [9]. A variety of methodologies have been employed to study rivers, including field measurements, numerical simulations, and laboratory experiments. Recently, big data and machine learning have become valuable tools, which benefit our understanding of river basin dynamics and supporting decision-making [10].

4. Objectives of this book

There is a need now more than ever to address the challenges of river basin management [11]. Studies predict that nearly one billion people reside in areas where action will likely be necessary, with approximately 365 million people living in basins where action is almost certain to be required [8]. The objectives of this book are to provide insights into tacking these challenges, predicting future changes, and developing adaptive strategies, such as flood prevention and disaster reduction, river channel restoration, maintaining river ecological balance, and ensuring sustainable water use. The chapters cover topics such as new approaches for water quality and environmental monitoring, soil erosion modeling, and the use of big data and machine learning in hydrology. Case studies on nature-based solutions for river restoration are also included. River basin management is a complex and urgent field. It requires a holistic approach to develop sustainable solutions for growing challenges. This book offers policymakers, researchers, water managers, and environmentalists practical knowledge and strategies to tackle these challenges effectively.

Author details

Chao Guo
Changjiang River Scientific Research Institute of Changjiang Water Resources Commission, Wuhan, Hubei, China

*Address all correspondence to: guoc@mail.crsri.cn

References

[1] Vörösmarty CJ, Mclntyre PB, Gessner MO, et al. Global threats to human water security and river biodiversity. Nature. 2010;**467**(7315):555-561

[2] Dethier EN, Renshaw CE, Magilligan FJ. Rapid changes to global river suspended sediment flux by humans. Science. 2022;**376**(6600):1447-1452

[3] Tollefson J. How hot will earth get by 2100? Nature. 2020;**580**:443-445

[4] IPCC. Climate Change 2021: The Physical Science Basis. Contribution of Working Group I to the Sixth Assessment Report of the Intergovernmental Panel on Climate Change; 2021

[5] William JR, Christopher WJ, Johan R, et al. The 2024 state of the climate report: Perilous times on planet earth. Bioscience. 2024;**74**(12):812-824

[6] Best J. Anthropogenic stresses on the world's big rivers. Nature Geoscience. 2019;**12**(1):7-21

[7] Grill G, Lehner B, Thieme M, et al. Mapping the world's free-flowing rivers. Nature. 2019;**569**(7755):215-221

[8] Palmer MA, Reidy Liermann CA, Nilsson C, et al. Climate change and the world's river basins: Anticipating management options. Frontiers in Ecology and the Environment. 2008;**6**(2):81-89

[9] Habersack H, Hein T, Stanica A, et al. Challenges of river basin management: Current status of, and prospects for, the river Danube from a river engineering perspective. Science of the Total Environment. 2016;**543**:828-845

[10] Ma K, He D, Liu S, et al. Novel time-lag informed deep learning framework for enhanced streamflow prediction and flood early warning in large-scale catchments. Journal of Hydrology. 2024;**631**:130841

[11] Falconer RA. Water security: Why we need global solutions. Engineering. 2022;**16**:13-15

Section 2

Monitoring Technologies for Eco-Environmental Parameters

Chapter 2

Electrical Conductivity Estimation in the Medina River, Texas, USA: An Integrated Approach Using Wavelet Analysis and Machine Learning Techniques

Salar Khani and Neda Khademi Shiraz

Abstract

Electrical conductivity (EC) is an important indicator for monitoring water quality in riverine systems. EC is inherently associated with the concentration of dissolved ionic compounds present in aqueous environments, including various salts and minerals. EC estimations are crucial for environmental monitoring and the overall health assessment of aquatic ecosystems. The present study investigated the application of discrete wavelet transform (DWT) in conjunction with artificial neural networks (ANNs) and multiple linear regression (MLR) models to predict daily river water EC. For this purpose, daily river discharge (Q) and EC time series from a hydrology station on the Medina River in San Antonio, Texas, USA, were used. DWT was used to decompose the daily data into several subseries. Then, to estimate one-day-ahead EC values, these subseries were introduced to the ANN and MLR models. To assess the prediction accuracy of the improved wavelet-neural network (WANN) and wavelet-regression (WR) models, EC estimation was also carried out using MLR and ANN models with the original data. Both the WANN and WR techniques outperformed single MLR and ANN methods. A comparison of the results indicated that the WR model had superior performance than the WANN, MLR, and ANN models for daily EC prediction. The R^2 values for the WR, WANN, MLR, and ANN models were 0.92, 0.87, 0.74, and 0.74, respectively. For the WR model, the root-mean-square error (RMSE) was 45.55, 46.08, and 25.19% less than those presented by the MLR, ANN, and WANN models, respectively. By the application of the WR method, an accurate daily EC estimator formula was obtained as well. The WR model also satisfactorily simulated the hysteresis in EC, demonstrating the effectiveness of wavelet analysis in extracting essential information embedded in original data.

Keywords: water quality, electrical conductivity, multiple linear regression, artificial neural networks, wavelet, hysteresis, Medina River

1. Introduction

Water quality deterioration impairs ecological function, puts human health at risk, and impedes socioeconomic development [1]. Monitoring water quality is crucial to maintaining the health of aquatic ecosystems [2], identifying pathogens, harmful chemicals, and other contaminants that pose risks to human health [3], and for resource management. Various water quality variables, including turbidity, pH, dissolved oxygen (DO), biological oxygen demand (BOD), total dissolved solids (TDS), and electrical conductivity (EC), are often used to monitor river water quality worldwide. These water quality indicators delineate a multifaceted phenomenon influenced by numerous hydrodynamic, hydrologic, and climate extremes (floods and droughts) [4] that function across an extensive spectrum of spatiotemporal domains [5].

Electrical conductivity, also called specific conductance or specific conductivity [6], is a key indicator of water quality, particularly in freshwater systems. Since it is predominantly influenced by total dissolved solids (TDS) and is intrinsically linked to the presence of dissolved ionic solutes in aqueous environments, it serves as a potential indicator of contaminants within surface water [7]. TDS and salinity levels of water can be estimated from its EC measurements. EC exhibits a positive correlation with both TDS and salinity, suggesting that elevated EC readings typically signify increased levels of TDS and salinity [8]. EC, which is reported in micromhos per centimeter (μmhos/cm) or microsiemens per centimeter (μS/cm) [9, 10], is a key parameter for evaluating water quality in many riverine systems and for assessing its appropriateness for diverse uses. Fluctuations in EC may indicate the presence of pollutants or contaminants within the water.

Monitoring of electrical conductivity is instrumental in recognizing prospective challenges, executing suitable water management strategies, and guaranteeing adherence to legislative criteria aimed at safeguarding both human health and ecological integrity [8]. However, a huge amount of water quality data, including EC, is obtained through continuous monitoring of rivers that needs to be analyzed to obtain valuable information such as trends and fast changes in the water quality parameter [11]. EC prediction is an important part of EC data analysis that should be performed before making any decision for water resources management. Through the prediction process, future EC levels can be estimated from historical data. The process is also beneficial for EC data gap filling and assessing the performance of the monitoring devices.

In the past two decades, various machine learning approaches such as artificial neural networks (ANNs), random forest (RF), and decision tree (DT) have been widely applied to analyze water quality time series. These models have been extensively used for predicting various water quality parameters, including EC, in rivers throughout the world. Ghorbani et al. [12] used multiple linear regression (MLR) and ANNs to predict EC in the Asi River in Antakya, Turkey. They concluded that EC estimations by the ANN models were more accurate than those obtained by the MLR model. Abozari et al. [13] employed an ANN model optimized with an imperialist competitive algorithm (ICA), particle swarm optimization (PSO), and genetic algorithm (GA) to predict EC in Gamasiab River, Iran. To predict the EC of the Babol-Rood River in northern Iran, Melesse et al. [14] applied different machine learning algorithms. Based on the correlation coefficient, they developed different input combinations from several water quality parameters and river discharge (Q) as inputs for the models and concluded that the developed machine learning models can predict EC with high accuracy.

A study for EC estimations in the Upper Indus River basin, Pakistan, was carried out by Shah et al. [15] using gene expression programming (GEP) and ANN models, and the GEP model outperformed the ANN model. In another study for predicting EC in this river, Alqahtani et al. [16] use GEP, ANN, and random forest (RF) models. The RF model was superior to both the ANN and GEP models. Kumar et al. [8] attempted to predict EC in the Upper Ganga basin using various machine learning techniques. The results of their research indicated that the RF method outperformed the co-adaptive neuro-fuzzy inference system (CANFIS), DT, and ANN models. Karbasi et al. [17] developed a convolutional neural network (CNN) model coupled with long short-term memory (LSTM) to estimate EC at two rivers in Australia. They concluded that their proposed model has a high capability to predict EC over the next 10 days.

Over the past decade, there has been a growing interest in developing hybrid algorithms to boost the prediction accuracy of machine learning approaches [18]. Most of the advancements have been made based on data preprocessing methods, including wavelet analysis. Recent studies confirm the capacity of wavelet transformation (WT) as a common tool for analyzing periodicities, trends, and fast changes within a time series [19–22]. Steel and Lange [23] utilized wavelet analysis to measure the water temperature variability across various temporal scales simultaneously. Ravansalar and Rajaee [24] used WT in conjunction with ANNs to estimate monthly EC levels of the Asi River in Turkey. Khani and Rajaee [25] developed two combination methods utilizing discrete wavelet transform (DWT) to estimate both the short- and long-interval DO levels at the Clackamas River, OR, USA. They concluded that DWT-based hybrid models increase the prediction accuracy of the water quality variables.

In this research, to predict daily electrical conductivity, two new hybrid models are developed based on the wavelet analysis, one the wavelet and artificial neural network (WANN) model, and the second the wavelet and multiple linear regression (WR) model. The objective of integrating wavelet analysis with the ANN and MLR models is to enhance the precision of EC predictions. The present study shows a new application of the WANN and WR conjunction models, which use decomposed subseries to predict river water EC and simulate its hysteresis. As a first study in this field, the capability of these models to predict daily EC was examined using various performance criteria. A straightforward explicit mathematical formula for the estimation of daily river water electrical conductivity was proposed using the WR model. Results obtained from this formula confirmed its applicability for predicting daily river water EC. The accuracy of the WR and WANN models in predicting daily EC and simulating its hysteresis were compared with those obtained from the single ANN and MLR models.

The remainder of this chapter is structured as follows: Section 2 outlines the study area and data; Section 3 covers MLR, ANN, and wavelet analysis; Section 4 involves model evaluation; and Section 5 discusses the application of the single MLR and ANN models, along with the hybrid WR and WANN models. The results from these models are discussed in Section 6, followed by a conclusion in Section 7.

2. Study area and EC data

2.1 Study area

To conduct this study, daily discharge (Q) and electrical conductivity (EC) data from station USGS Gage 08180700 (latitude 29° 20′ 5.00″, longitude 98° 41′ 22.00″) on the Medina River at San Antonio, Texas, in the USA were used. Data available from the U.S. Geological Survey, National Geospatial Program. The daily data used in this research covered 3 years (1096 days) from 2008 to 2010. The recorded data spans calendar years, with the starting month being January 2008 and the ending month

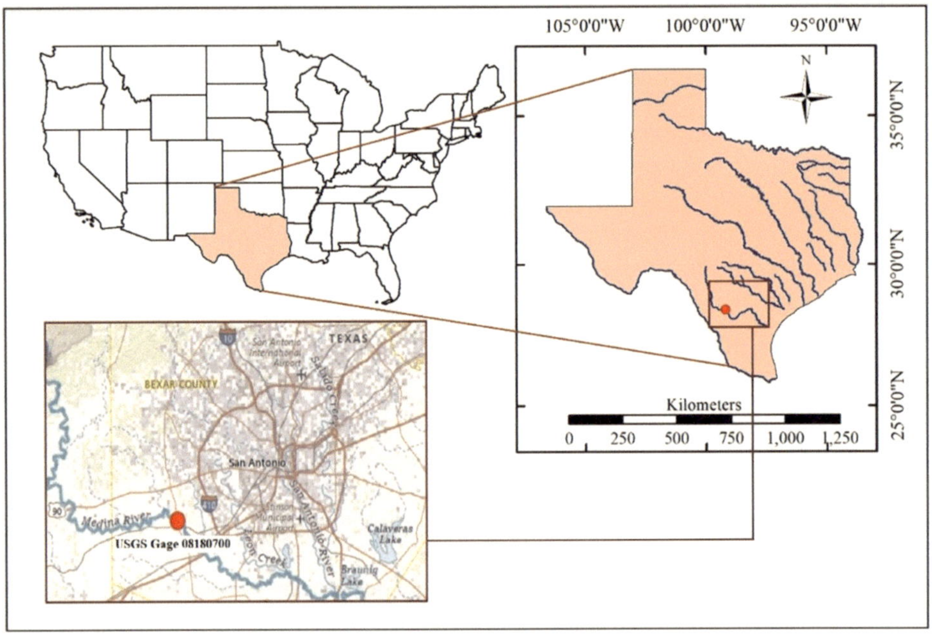

Figure 1.
Medina River and the location of the USGS monitoring station (the map was created using ArcGIS software, version 10.8.1: https://support.esri.com/zh-cn/products/desktop/arcgis-desktop/arcmap/10-8-1), USA shape file, and the base map of San Antonio is from https://www.usgs.gov, and the map of Texas Rivers is from https://www.twdb.texas.gov/mapping/gisdata.asp.

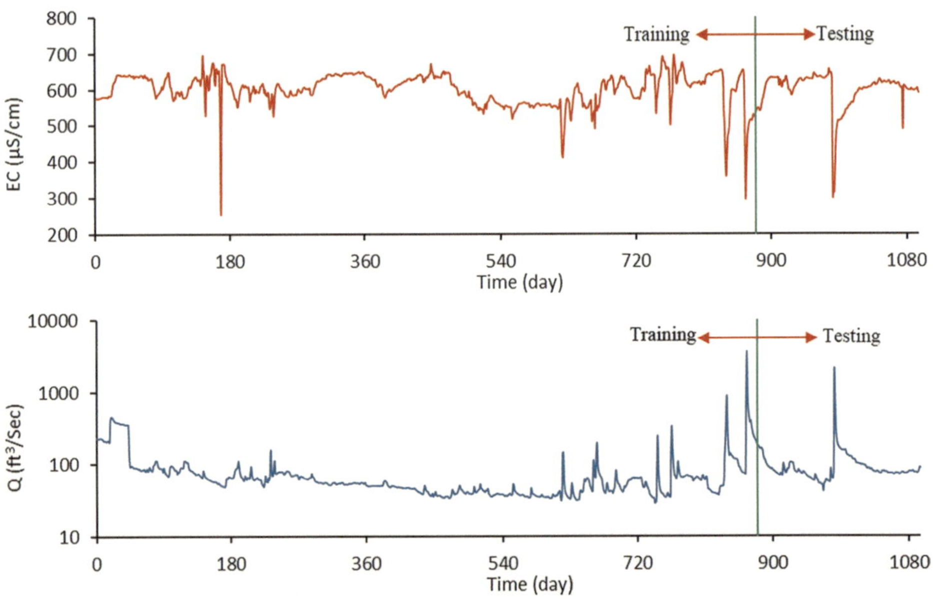

Figure 2.
Daily EC and Q time series for the monitoring station (Q time series are presented on a semi-logarithmic scale, and its values, when divided by 35.315, are expressed in cubic meters per second (m^3/s)).

being December 2010. The Medina River and the location of the selected monitoring station are presented in **Figure 1**.

The Medina River, which passes through the west of San Antonio, is a main supply to recharge the Edwards Aquifer. This aquifer serves as the primary source of drinking water for San Antonio and its neighboring regions, providing for approximately 2.3 million residents, based on the information available on the Edwards Aquifer Authority webpage. The primary water quality concern in South-Central Texas is the risk of contamination of the Edwards Aquifer [26]. Therefore, protecting the Medina River from pollutants is a major issue. The Medina River water quality, especially in this region, is affected by agricultural activities.

From the 1096 daily EC and river discharge data, the first 880 data points (80%) were considered for training the models, and the other 216 data points (20%) were used to evaluate the performances of the models. The EC and river discharge time series for the investigation period are shown in **Figure 2**. Since the variation between the highest and lowest observed discharge values of this river during the study period was significant, the Q time series is presented in a semi-logarithmic scale.

2.2 Statistical analysis

The results of the statistical analysis of daily river discharge (Q) and electrical conductivity (EC) data are summarized in **Table 1**. For the train subset, test subset, and the entire data set, the minimum, maximum, mean, standard deviation (S_d), skewness coefficient (C_{sx}), and the lags 1 day to 4 days autocorrelation coefficients (R_1, R_2, R_3, and R_4) were calculated. Similar to all empirical methods, ANN and MLR techniques are more effective when they operate within the data range used for training rather than extrapolating beyond it [27]. In other words, when splitting the entire data into the train and test data, the maximum and minimum measurements should happen in the train subset. In such a situation, a higher accuracy in the prediction values of the model may be expected [28].

From **Table 1**, it is evident that the maximum and minimum EC and Q are included in the training subset. Also, it can be seen that both the training and testing subsets exhibit fairly comparable statistical characteristics. EC autocorrelation coefficients,

	All data		Train subset		Test subset	
	Q (ft^3/sec)	EC (μS/cm)	Q (ft^3/sec)	EC (μS/cm)	Q (ft^3/sec)	EC (μS/cm)
Mean	86.28	597.89	82.37	598.49	102.21	594.45
Min	28.10	251	28.10	251	41.30	299
Max	3563.90	694	3563.9	694	1943.70	654
S_d	144.84	45.94	144.93	44.77	143.39	50.37
C_{sx}	15.49	−2.22	16.68	−1.86	10.93	−3.215
R_1	0.538	0.872	0.555	0.875	0.456	0.861
R_2	0.317	0.725	0.353	0.718	0.157	0.743
R_3	0.265	0.616	0.303	0.618	0.097	0.610
R_4	0.241	0.538	0.276	0.546	0.080	0.510

Table 1.
Results of the statistical analysis of the data.

especially R_1 and R_2, were considerable, but Q autocorrelation coefficients, especially R_2, R_3, and R_4, were very low for both the train and test subsets. For both the subsets, the skewness of the EC and Q data was low, which is appropriate for EC estimation.

In the modeling procedure, to assure high prediction accuracy by the models, it is essential to use an appropriate set of input data. In other words, choosing an appropriate input combination is a very important step in predicting river water quality. There are many different procedures for input selection for the predictive models. In this research, the correlation coefficients (R) were considered to produce different input combinations for the models. For this purpose, using Eq. (1), the correlation coefficients between the measured ECt and Q_{t-i} (i = 0, 1, 2, 3, and 4) time series are computed to extract the dependency of EC on river discharge. The calculated correlation coefficients are summarized in **Table 2**.

$$R = \frac{\sum (Q_i - Q_{mean})(EC_i - EC_{mean})}{\sqrt{\sum (Q_i - Q_{mean})^2 \sum (EC_i - EC_{mean})^2}} \tag{1}$$

where EC_{mean}, EC_{min}, and EC_{max} are the mean, minimum, and maximum EC, and Q_{mean}, Q_{min}, and Q_{max} are the mean, minimum, and maximum Q, respectively.

Greater absolute values of correlation coefficients, close to 1, signify a stronger relationship between the parameters. The minus sign behind the correlation values presented in **Table 2** indicates that river water electrical conductivity negatively depends on river water discharge. It means that an increase in river water discharge leads to a decrease in river water EC values. Also, higher EC values are expected in dry periods when river discharge is low. As it is shown in **Table 2**, the correlations between EC_t and Q_t, Q_{t-1}, and Q_{t-2} are relatively high. Therefore, in this study, to estimate EC values, Q_t, Q_{t-1}, and Q_{t-2} time series were also considered as the input variables to the predictive models.

Data normalization is an essential step that should be carried out before introducing data to the models. This is due to the use of preprocessed non-dimensional data and ensuring that all variables receive equal consideration during the model calibration and testing process. Various techniques for normalizing water quality data are summarized in [18]. In our research, we used the following simple linear mapping equation to normalize the data between 0 and 1.

$$EC_{normalized} = \frac{(EC_i - EC_{min})}{(EC_{max} - EC_{min})} \tag{2}$$

where EC_{min} and EC_{max} are the minimum and maximum EC measurements, respectively.

Time series	All data	Train subset	Test subset
Q_t	−0.407	−0.328	−0.693
Q_{t-1}	−0.380	−0.307	−0.645
Q_{t-2}	−0.312	−0.233	−0.601
Q_{t-3}	−0.234	−0.169	−0.467
Q_{t-4}	−0.179	−0.128	−0.361

Table 2.
The calculated correlation coefficient between EC and Q.

3. Models description

3.1 Artificial neural networks (ANNs)

Artificial Neural Networks are computational algorithms that replicate the neural organization of the human brain, specifically designed for pattern recognition, and perform various tasks, such as classification and regression [29, 30]. Because of their high capability of learning and modeling complex patterns within the data, these models have been extensively applied as predictive models in various fields. A basic architecture of ANNs is the three-layer feed-forward neural network, which includes an input layer, a single hidden layer, and an output layer, where the flow of information is unidirectional without feedback loops [31]. Each of the layers contains some computation nodes, also called neurons, and each neuron in the network sums its weighted inputs and utilizes a nonlinear activation function [32]. When there is no feedback from the outputs of the neurons to the inputs in the network, it is called a feed-forward neural network [33]. The backpropagation algorithm is fundamental for training ANNs, allowing the model to minimize error through a two-phase process: the forward pass, where inputs generate outputs, and the backward pass, where gradients of error are computed and used to update weights [34]. This iterative tweaking of weights enables the neural network to learn from its mistakes and improve performance on predictive tasks [35]. The feed-forward backpropagation (FFBP) is among the widely used neural networks applied for water quality predictions worldwide [18, 36].

In this study, the Levenberg-Marquardt (LM) [37] optimization algorithm, which surpasses simple gradient descent and various conjugate gradient methods across a broad range of problems [38], was used to train the FFBP networks. This optimization method is an iterative approach designed to find the lowest value of a function represented as the sum of squares of nonlinear functions [39].

A mathematical expression of a neural network with three layers, comprising I, J, and one node in its input, hidden, and output layers, respectively, is presented in Eq. (3) [40, 41].

$$Y = f_1\left(\sum_{j=1}^{J} w_j^o f_2\left(\sum_{i=1}^{I} w_{ij}^h X_{pi} + b_1\right) + b_2\right) \tag{3}$$

where Y is the network's output, X_{pi} is the input data, w_{ij}^h indicates the weight of the connection between the ith node in the input layer and the jth node of the hidden layer, w_j^o refers to the weight of the link between the jth node in the hidden layer and the output layer node [42]. Additionally, b_1 and b_2 are bias terms, and $f_1()$ and $f_2()$ are respectively the hidden and output layers' activation functions.

3.2 Multiple linear regression (MLR)

Multiple Linear Regression (MLR) is a conventional procedure for modeling the linear connection between a dependent parameter and one or more independent parameters [43]. Its simplicity makes it relatively easy to implement. However, it does have limitations in forecasting, particularly when dealing with complex dynamics and a large volume of noisy data [44]. The common form of MLR is given in Eq. (4).

$$Y = a_0 + \sum a_i X_i \tag{4}$$

In this equation, a_0 is the intercept, a_i is the regression coefficient of the descriptor X_i, and Y is the model estimation.

3.3 Wavelet analysis

Wavelet analysis has developed into an effective method for interpreting data across multiple fields, such as signal processing, image analysis, and financial forecasting. In contrast to conventional Fourier analysis, which decomposes signals into sine and cosine components, wavelet analysis allows for the representation of data at various scales and resolutions. This flexibility makes wavelets especially useful for examining transient and non-stationary signals, facilitating the retrieval of localized information that traditional methods may miss [45]. Wavelet analyses are performed using wavelet functions, also called mother wavelets. The mathematical presentation of the wavelet function is provided in Eq. (5).

$$\psi(t) = \frac{1}{\sqrt{s}} \psi\left(\frac{t-\tau}{s}\right) \tag{5}$$

where s and τ are the scaling and translation parameters, respectively.

The mother wavelet $\psi(t)$ acts as the foundation from which all other wavelet functions are derived via scaling and translation [46]. This scaling process enables the analysis of signals across various frequency ranges, allowing wavelets to effectively capture localized characteristics in time-frequency representations.

The Discrete Wavelet Transform (DWT) is a mathematical technique implemented to analyze and break down discrete signals into approximations and detail coefficients at multiple levels. The approximation coefficients reflect low-frequency information, while the details capture high-frequency variations. Applying DWT, both the time and frequency information of the signal can be extracted. More information on the DWT is provided by Shensa [47].

The wavelet transformation of the signal $f(t)$ mathematically can be expressed as:

$$W_{\psi}f(s,\tau) = \frac{1}{\sqrt{s}} \int_R f(t)\psi\left(\frac{t-\tau}{s}\right) dt \tag{6}$$

The wavelet transform involves breaking down $f(t)$ at various resolution levels. In practical applications, the successive wavelet is typically discrete, and for a discrete time series, the DWT can be expressed as [48]:

$$W_{\psi}f(j,k) = 2^{-j/2} \sum_{i=0}^{N-1} f(t)\psi\left(2^{-j}t - k\right) dt \tag{7}$$

where, $W_{\psi}f(j,k)$ represents the coefficient associated with a discrete wavelet applied at a specific scale level $s = 2j$ and translation $\tau = 2jk$.

The first step in conducting wavelet analysis involves selecting a suitable wavelet function. Different wavelet families, including Haar, Simlet, and Daubechies, have been used for time series decomposition or noise removal from the data. The

Daubechies wavelet family is widely utilized due to its compact support and orthogonality properties [46]. Another critical aspect of the process involves identifying the optimal degree of decomposition. The decomposition level indicates how many times the signal should be divided into its approximation and detail components.

During wavelet analysis, the original signal decomposes into one approximation (A) and one or more details (Ds) coefficients using low and high-pass filters [49]. The approximation component represents the trend in the time series, and the details comprise its high frequencies. In the present study, the A and Ds subseries of the original EC and Q time series are obtained by applying DWT. The Daubechies-2 (db2) wavelet function was utilized to break down the time series at three levels.

4. Model evaluation metrics

The root-mean-square error (RMSE), the mean absolute error (MAE), and the Nash-Sutcliffe coefficient of efficiency (NS) performance measures (Eqs. 8–10) were used to assess the accuracies of the models for EC estimation. In these equations, n denotes the number of observations in the dataset. The models achieve optimal predictions when NS, MAE, and RMSE are close to 1, 0, and 0, respectively. Detailed information on NS evaluation metrics is available in [50].

$$NS = 1 - \frac{\sum_{i=1}^{n}\left(EC_{i(observed)} - EC_{i(predicted)}\right)^2}{\sum_{i=1}^{n}\left(EC_{i(observed)} - EC_{mean}\right)^2} \tag{8}$$

$$RMSE = \sqrt{\frac{\sum_{i=1}^{n}\left(EC_{i(observed)} - EC_{i(predicted)}\right)^2}{n}} \tag{9}$$

$$MAE = \frac{\sum_{i=1}^{n}\left|EC_{i(observed)} - EC_{i(predicted)}\right|}{n} \tag{10}$$

5. Model implementations

5.1 ANN and MLR models

Based on the statistical analysis presented in **Table 1** and the correlations between river electrical conductivity and discharge time series shown in **Table 2**, the six different combinations, which consist of various values of EC and Q, are considered as input data for the ANN and MLR models to predict one-day-ahead EC (EC_{t+1}). The generated input combinations for the prediction of EC_{t+1} are shown in **Table 3**:

In this research, an FFBP neural network with three layers, comprising one input layer, one hidden layer, and one output layer in its structure, was implemented. The Levenberg-Marquardt method was used for training the ANN models. In ANN models, choosing the right activation function for the nodes is crucial.

Typically, activation functions in neural networks possess three key properties: they are bounded, continuous, and non-constant. Among these, the Sigmoid function is the most commonly used [25]. In the current study, the Sigmoid and Purlin functions were employed as activation functions for the hidden and output layers, respectively, to enhance the effectiveness of the neural network. The optimal number of

Input combination	EC time series			Q time series	
C1	EC_t				
C2	EC_t	EC_{t-1}			
C3	EC_t	EC_{t-1}	EC_{t-2}		
C4	EC_t			Q_t	
C5	EC_t	EC_{t-1}		Q_t	
C6	EC_t	EC_{t-1}		Q_t	Q_{t-1}

Table 3.
The time series combinations of the inputs for the ANN and MLR models.

neurons in the hidden layer was established through iteration. The developed ANN models were trained using the train dataset and then assessed using the test data.

In addition to the ANN model, MLR was also used to create the relation between the input variables and EC for all the six input combinations shown in **Table 3**. By applying the MLR model to each input combination, six different regression equations were created. These equations were obtained based on the training data set. Subsequently, the model's predictive capabilities were assessed using the same datasets that were used to test the ANN models.

5.2 Wavelet-ANN combination model

This study presents a new application of the WANN combination model for the prediction of river electrical conductivity. The goal of integrating the discrete wavelet transform with artificial neural networks is to enhance the accuracy of the predicted EC values. The WANN combination model uses decomposed subseries as inputs to the ANN model. To perform this research, the Daubechies-2 (db2) wavelet function was used, and the EC and Q time series were decomposed at three levels utilizing discrete wavelet transform. During the decomposition process, the original time series is initially divided into an approximation (A1) and a details (D1) subseries. This process can be extended, allowing the approximation subseries to be further decomposed into lower-resolution components. In the WANN model, the approximation and details components of the EC and Q time series served as input data for the ANN models. The structure of the WANN hybrid model for prediction of 1-day-ahead EC values is presented in **Figure 3**.

5.3 Wavelet-regression combination model

In this research, the wavelet-regression (WR) combination model is implemented to predict river electrical conductivity. This model is created by integrating two approaches: discrete wavelet transformation and multiple linear regression. It is anticipated that utilizing the approximation and details components of the time series will enhance the predictability of the regression model. The same DWT-obtained components of EC and Q time series, previously used in the development of the WANN model, were considered inputs for the WR model to predict one-day-ahead EC values. **Figure 3** represents the structure of the proposed wavelet-regression model.

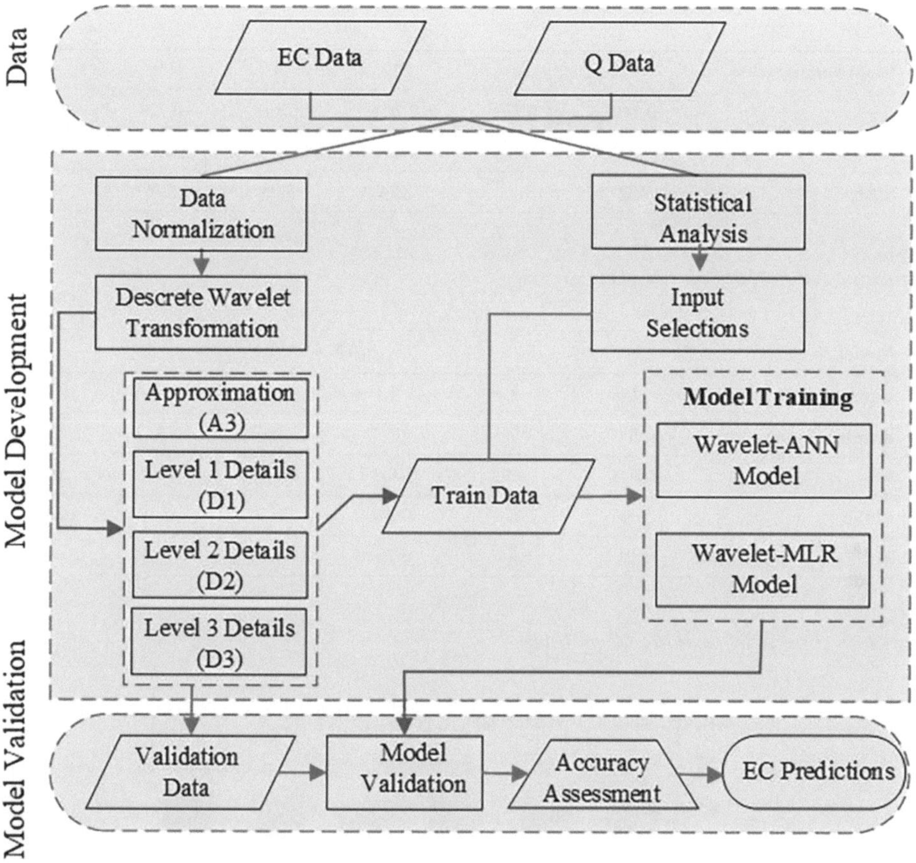

Figure 3.
WR and WANN modeling diagram for EC prediction.

6. Results and discussion

The prediction of one-day-ahead EC is performed using MLR and ANN models. The performances of these models for the six input combinations are summarized in **Tables 4** and **5**, respectively. As can be seen from these tables, the higher accuracies of the MLR and ANN models obtained from the C4 (EC_t and Q_t) input combination, which includes observed values of current day EC and Q. For C4, the ANN and MLR models predicted the electrical conductivity with NS = 0.737 and 0.742, RMSE = 25.78 and 25.53, and MAE = 7.08 and 7.25, respectively. Results also indicate that for the input combination C1, which uses only the current day EC time series (EC_t), the accuracy of the ANN and MLR models is higher than all other input combinations except the C4. As indicated earlier, the optimum number of neurons in the hidden layer was identified based on a trial-and-error approach. In **Table 5**, for all the six input combinations, the optimum number of neurons in the hidden layer of the ANN models is presented as well. For the most accurate ANN model, the ANN architecture is represented as ANN (2, 4, 1), indicating that this model has two neurons in its input layer, 4 neurons in its hidden layer, and one neuron in its output layer.

A comparison between different ANN and MLR models based on the RMSE values is presented in **Figure 4**. From this figure, it can be seen clearly that, for each input

Model	MLR					
Input combination	**1**	**2**	**3**	**4**	**5**	**6**
NS	0.740	0.729	0.704	0.742	0.728	0.728
RMSE	25.60	26.16	27.34	25.53	26.18	26.18
MAE	7.32	7.78	7.84	7.25	7.76	7.76

Table 4.
Evaluation of the MLR models in EC prediction.

Model	ANN					
Input combination	**1**	**2**	**3**	**4**	**5**	**6**
neurons	5	5	7	4	3	8
NS	0.733	0.697	0.703	0.737	0.705	0.715
RMSE	25.96	27.64	27.40	25.78	27.29	26.81
MAE	6.81	6.84	6.72	7.08	8.96	7.01

Table 5.
Evaluation of the ANN models in EC prediction.

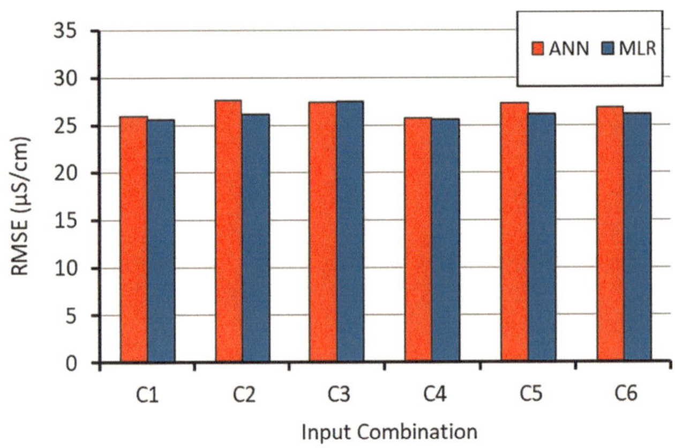

Figure 4.
Comparison of RMSE values obtained from the ANN and MLR models for the six input combinations.

combination, the MLR model outperforms the corresponding ANN model to a slight degree. The most accurate MLR equation, obtained for the input combination C4, is presented in Eq. (11). In this equation, the minus sign behind the Q_t, as discussed before, indicates that river discharge has a reverse impact on daily EC values. The best accurate ANN and MLR models performed with NS = 0.737 and 0.742, RMSE = 25.78 and 25.53, and MAE = 7.08 and 7.25, respectively.

$$EC_{t+1} = 0.889EC_t - 0.007Q_t + 78.79 \tag{11}$$

The hybrid WANN and WR models were employed to enhance EC prediction accuracy. The calculated performance measures for the WR and WANN models are summarized in **Table 6**. As mentioned earlier, both the MLR and ANN demonstrated

Model	WANN							WR
neurons	5	9	10	11	12	15	19	—
NS	0.642	0.682	0.858	0.863	0.713	0.808	0.743	0.923
RMSE	30.03	28.31	18.89	18.58	26.91	22.00	25.45	13.90
MAE	15.54	8.68	7.04	6.45	8.68	9.87	9.08	5.97

Table 6.
Performances of the WR and WANN models for predicting EC_{t+1}.

their highest level of accuracy for the input combination C4. It means that for Medina River, in the hydrological station selected for this study, using observed values of current day Q and EC would result in the best prediction for one-day-ahead electrical conductivity. Therefore, the prediction of EC with the WANN and WR models was conducted by applying the DWT subseries of the EC_t and Q_t. The decomposed subseries of EC at three levels is presented in **Figure 5**. Similar subseries of the Q data were also obtained and used for the development of the hybrid models.

The performance of the WANN model for various numbers of neurons in the hidden layer of its ANN structure is shown in **Table 6**. This table indicates that the WANN with 11 neurons in the hidden layer of its ANN structure provides the most accurate EC predictions. The NS, RMSE, and MAE evaluation metrics for the best WANN model were 0.863, 18.58, and 6.45, respectively. The results of the WR model are presented in **Table 6** as well. For this model, the NS, RMSE, and MAE statistics were 0.923, 13.90, and 5.97, respectively.

The application of the WR model results in a straightforward equation that predicts EC_{t+1} with considerable accuracy (see Eq. (12)). This equation is a simple multiple linear regression that estimates the river electrical conductivity time series from the DWT-obtained subseries of the original time series of river discharge and electrical conductivity.

$$\begin{aligned} EC_{t+1} &= 15.291 - 0.915D1_{EC_t} + 0.014D1_{Q_t} + 0.601D2_{EC_t} + 0.010D2_{Q_t} \\ &+0.717D3_{EC_t} - 0.079D3_{Q_t} + 0.977A3_{EC_t} - 0.017A3_{Q_t} \end{aligned} \tag{12}$$

Comparison between the prediction accuracy of the hybrid WANN and WR model and the single ANN and MLR models revealed that the decomposition of input time series significantly improved the EC predictions. From **Tables 4–6** it is clear that both the WANN and WR models outperformed the ANN and MLR models. The accuracy of the WR model was superior compared to the three other models. The results of the most accurate MLR, ANN, WANN, and WR models in terms of RMSE are presented in **Figure 6**. The comparison of results reveals that the root-mean-square errors for the WR model were 45.55, 46.08, and 25.19% less than the MLR, ANN, and WANN models, respectively.

The scatter plots of the predicted EC by the models versus the corresponding observed EC during the test period are presented in **Figure 7**. It can be seen that both the ANN and MLR overestimated the minimum EC values. While the WANN model presented underestimations of the minimum EC, predictions of the WR model were close to the 1–1 line of the scatter plots. The EC predictions by the WR model were also more accurate than the three other models during high EC periods. **Figure 7** also

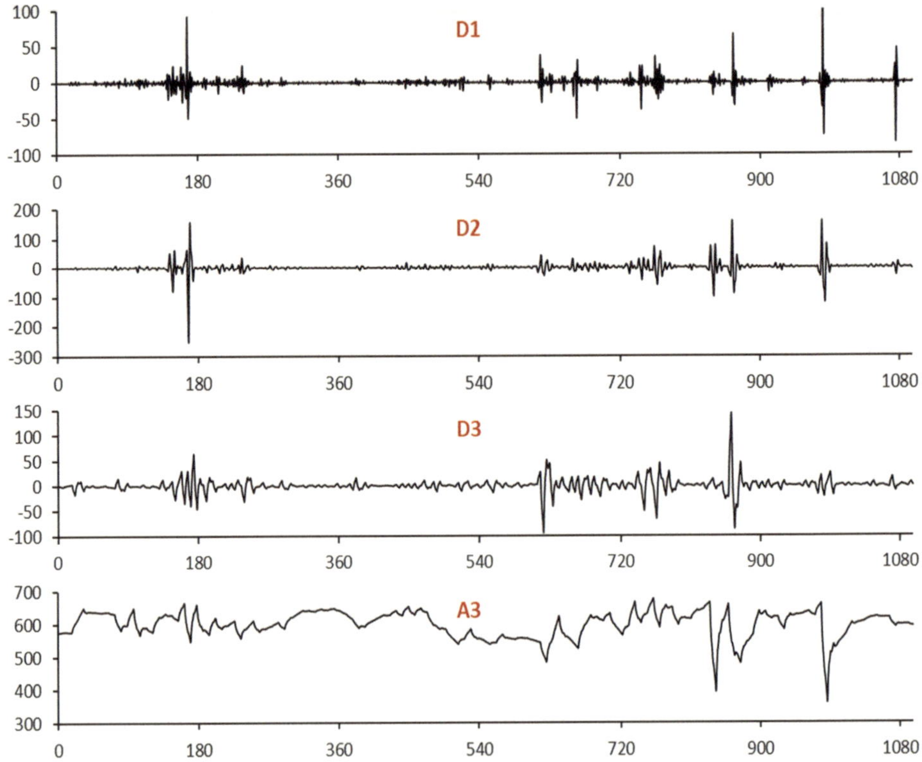

Figure 5.
Decomposed subseries of the EC data at three levels.

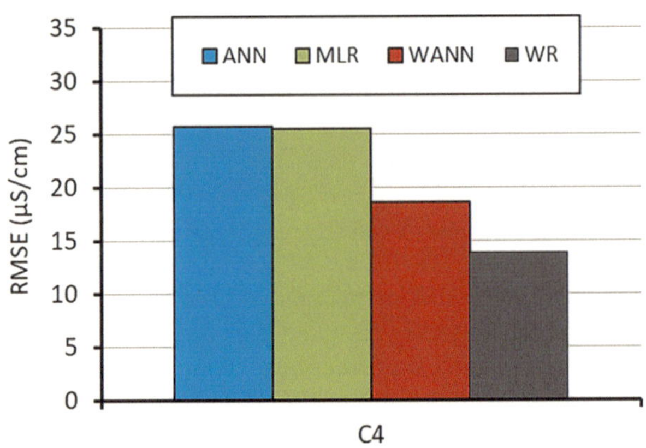

Figure 6.
Comparison between the RMSE of the best-performing ANN, MLR, WANN, and WR models.

provides the linear trendline and their corresponding equations and coefficient of determination (R^2) values for the predictive models. The higher R^2 value of the WR and WANN models is another indicator confirming their superiority to the single ANN and MLR models.

The residuals of the EC predictions by the models are shown in **Figure 8**. From this figure, the underestimations of the low EC levels by the WANN model and the over-estimations of the other models, especially the ANN and MLR models, are observable.

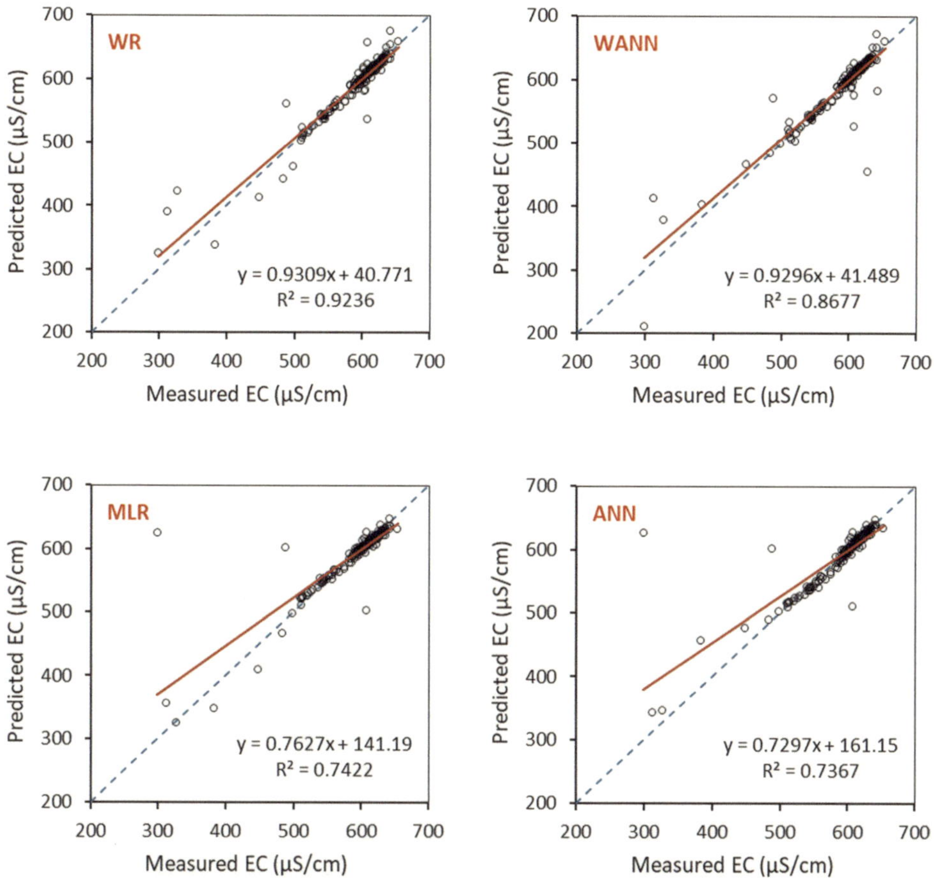

Figure 7.
Scatter plots of the predicted EC by the models versus measured EC.

In the low EC period, the residuals of the WR model were lower than those for the three other models.

From the results of both hybrid models applied in this research, it was evident that wavelet analysis played a crucial role in analyzing the EC and Q time series, allowing for the extraction of significant features within the data. This study demonstrated that the WR model is an effective method for predicting river electrical conductivity, thanks to its superior accuracy. Additionally, the use of WR models is straightforward compared to WANN and ANN models, providing fast and more precise predictions for water quality.

The observed time series of EC and Q shows an inverse dependency of the river water electrical conductivity to the river discharge amount. In other terms, the river displays hysteresis in its EC levels. This hysteresis phenomenon can also be seen in other hydrological processes, such as river suspended sediment concentrations [51] and river dissolved oxygen concentrations [25]. The hysteresis phenomenon as a nonlinear signature of the process is usually exhibited at different time and space scales [52]. In the following, using the WANN and WR models, an attempt was made to detect the hysteresis in EC.

The high river discharge of July 11th, 2010 (**Figure 9(A)**) and September 8th, 2010 (**Figure 9(B)**) were selected for the hysteresis analysis of EC. For the two events, the hydrographs and observed EC series are presented in the left side of **Figure 9**. In the

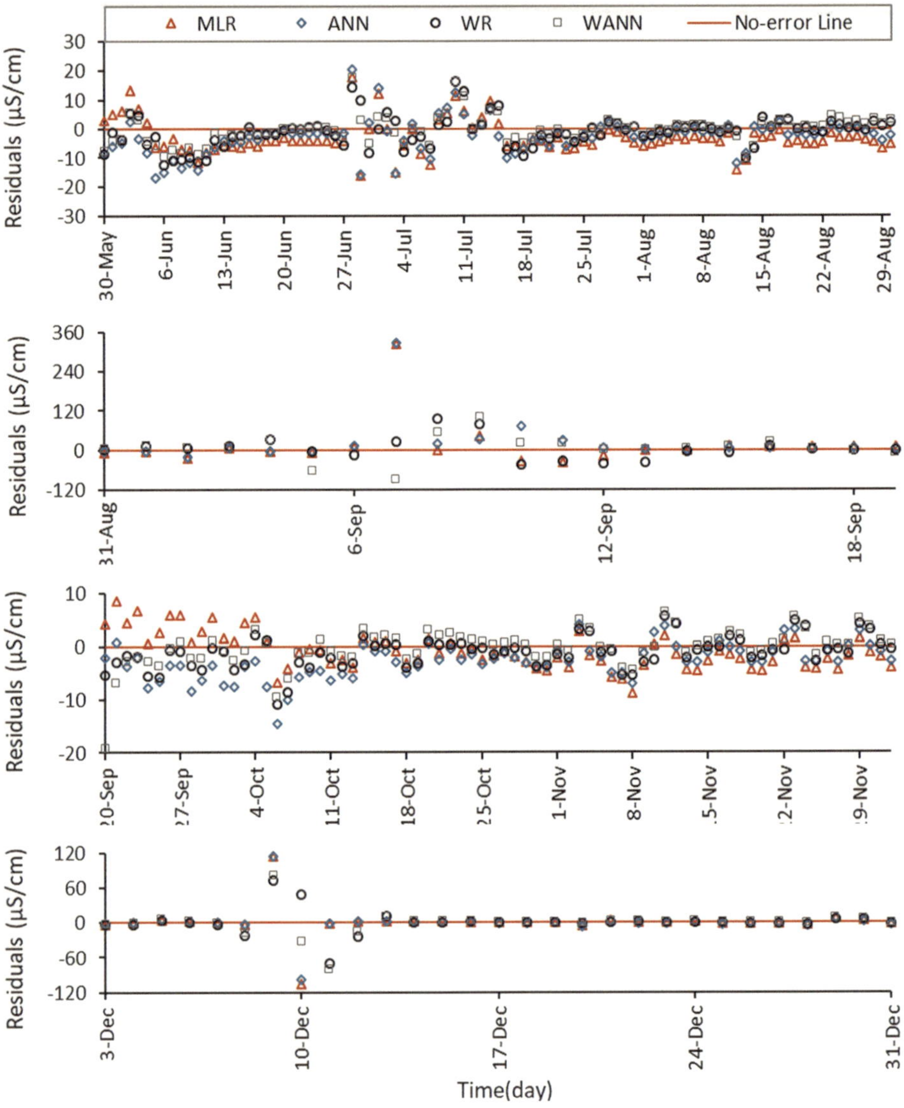

Figure 8.
EC prediction residuals by the models at different time intervals for the test period.

right side of this figure, the actual EC-Q loops and the corresponding simulated loops by the WR, WANN, MLR, and ANN models are shown. These EC-Q graphs were used to examine the capability of the models in simulating EC hysteresis loops. For the period A from July 7th, 2010 to July 21st, 2010, in which Q has relatively low and smooth variations, all the models were able to mimic the EC-Q loops.

For the period B from September 7th, 2010 to September 12th, 2010, river discharge rapidly increased from 61 ft^3/sec to 1943 ft^3/sec, resulting in a decrease in EC from 642 µS/cm to 299 µS/cm. Following these extreme values within this period, Q decreases to 193 ft^3/sec, and EC reaches its local maximum value of 483 µS/cm. For this period, as shown in **Figure 9(B)**, the ANN and MLR models failed to detect

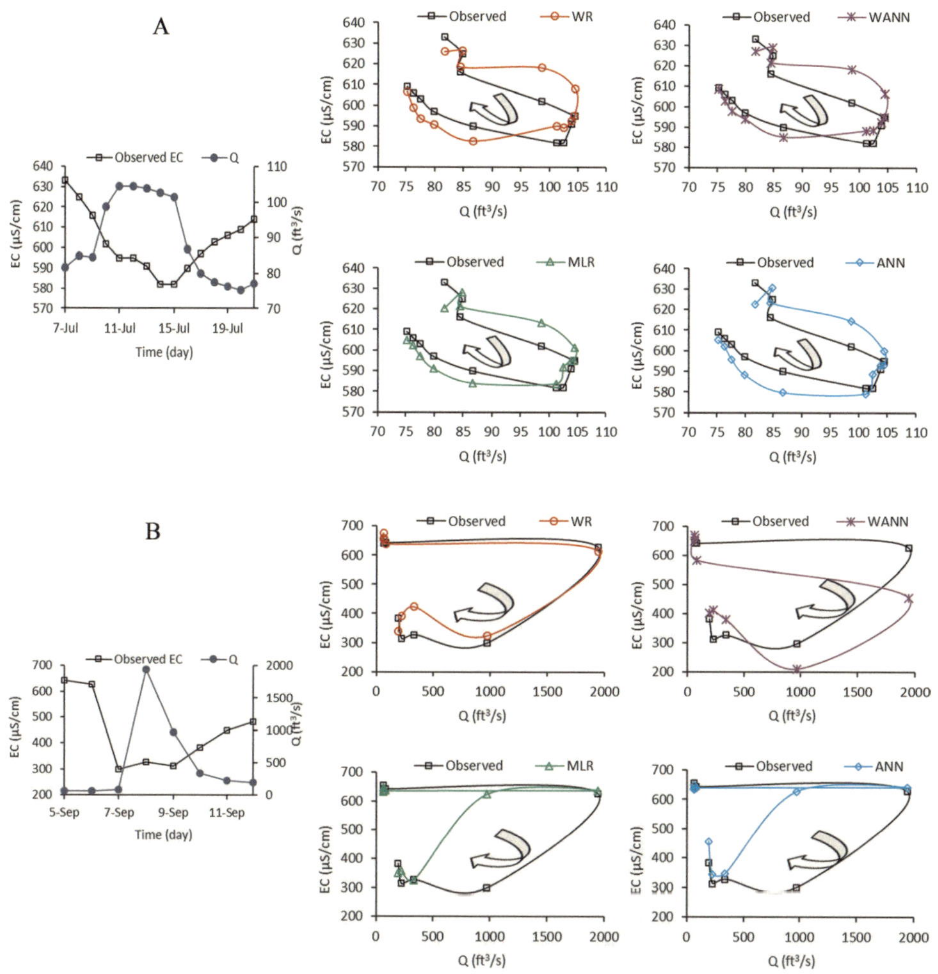

Figure 9.
(A) and (B) Time series of the river discharge, observed EC, and corresponding simulated EC-Q hysteresis loops during the two periods.

hysteresis within the EC time series. However, the WR and WANN models satisfactorily followed the clockwise (positive) hysteresis loop during the high discharge levels. It can be seen that the EC-Q loop simulated by the WR model is in good agreement with the actual EC-Q graph. In general, the wavelet-based models simulated hysteresis in EC better than the ANN and MLR models, confirming the capability of the DWT for studying EC-Q relationships in the river system.

7. Conclusions

The present study attempted to examine the performance of the discrete wavelet transform (DWT) in conjunction with ANN and MLR models for predicting river water electrical conductivity. For this purpose, two wavelet-ANN and wavelet-regression combination models were developed. The procedure involved decomposing EC and discharge data using the Daubechies-2 wavelet function across three levels, generating one approximation and three detail components. These decomposed subseries were then employed as inputs for the ANN and MLR models.

One-day-ahead EC levels were also predicted using a three-layer feed-forward neural network and simple MLR model. The results indicated that the MLR models slightly outperformed the ANN models. Although the WANN model, by employing the artificial neural network technique in its structure, provided better performance than the ANN and MLR models, its accuracy was lower than the WR model. The WR model demonstrated superior accuracy in predicting EC levels and was closely aligned with the measured values across various ranges. The optimal input combination for predicting one-day-ahead EC was identified as the current day's EC and river discharge values.

In this study, EC hysteresis simulations were carried out to detect the EC-Q relationship and simulate EC variations during the river's high- and low-flow conditions. The WR model excelled in simulating EC hysteresis, reflecting dynamic variations during high- and low-flow periods. In contrast, the WANN, ANN, and MLR models were unable to replicate this phenomenon.

The application of wavelet analysis resulted in a reasonable improvement in the prediction accuracy of EC and simulation of its hysteresis. This is because of its unique ability to concurrently analyze all temporal scales. The applied WR model offers a straightforward mathematical structure, significantly less complex than the WANN and ANN models. This simplicity allows the WR model to effectively model daily EC. In contrast, the WANN and ANN models are classified as black-box models, meaning that the relationships between the input and output data are unknown. The EC of any river may depend on many known and unknown hydrologic, land use, and physical parameters. In this study, the models were developed using a long period of daily EC and river discharge data. Therefore, the simple formula obtained from the WR model can be used as a practical tool to accurately predict daily EC levels in the Medina River.

Previous studies primarily focused on predicting monthly EC values using monthly data as inputs. These models incorporated various water quality, meteorological, and hydrological parameters, applying an array of machine learning techniques to estimate monthly EC levels. In contrast, the current chapter presents a novel approach by utilizing daily EC and river discharge data to estimate one-day-ahead EC levels. This shift from monthly to daily estimation is significant, as it allows for enhanced responsiveness to dynamic river conditions. We compared the performance of the proposed WR model against those from prior studies using RMSE metrics. The RMSE by the model was lower than numerous established models, including MLR, MNLR, ANN, GA, PSO, RF, DT, and CANFIS. This comparison underscores the superior accuracy of the proposed WR model for daily EC estimation in river water, highlighting its potential for more effective and timely water quality management.

Author details

Salar Khani[1]* and Neda Khademi Shiraz[2]

1 University of Tehran, Tehran, Iran

2 Tarbiat Modares University, Tehran, Iran

*Address all correspondence to: salar.khani@gmail.com

References

[1] du Plessis A. Persistent degradation: Global water quality challenges and required actions. One Earth. 2022;**5**(2): 129-131

[2] Carpenter SR, Caraco NF, Correll DL, Howarth RW, Sharpley AN, Smith VH. Nonpoint pollution of surface waters with phosphorus and nitrogen. Ecological Applications. 1998;**8**(3): 559-568

[3] World Health Organization. Guidelines for Drinking-Water Quality. Geneva: World Health Organization; 2021

[4] Fabian PS, Kwon HH, Vithanage M, Lee JH. Modeling, challenges, and strategies for understanding impacts of climate extremes (droughts and floods) on water quality in Asia: A review. Environmental Research. 2023;**225**: 115617

[5] Najah A, El-Shafie AH, Karim OA, Jaafar O, El-Shafie AH. An application of different artificial intelligences techniques for water quality prediction. International Journal of Physical Sciences. 2011;**6**(22):5298-5308

[6] Weiner ER. Applications of Environmental Aquatic Chemistry: A Practical Guide. Boca Raton: CRC Press; 2008

[7] Ahmadianfar I, Jamei M, Chu X. A novel hybrid wavelet-locally weighted linear regression (W-LWLR) model for electrical conductivity (EC) prediction in surface water. Journal of Contaminant Hydrology. 2020;**232**:103641

[8] Kumar D, Singh VK, Abed SA, Tripathi VK, Gupta S, Al-Ansari N, et al. Multi-ahead electrical conductivity forecasting of surface water based on machine learning algorithms. Applied Water Science. 2023;**13**(10):192

[9] Spellman FR. The Science of Water: Concepts and Applications. Boca Raton: CRC Press; 2020

[10] Heydari M, Olyaie E, Mohebzadeh H, Kisi Ö. Development of a neural network technique for prediction of water quality parameters in the Delaware River, Pennsylvania. Middle-East Journal of Scientific Research. 2013;**10**:1367-1376

[11] González PS, Stehr A, Barra RO. Assessment of water quality trends through the application of an aggregated water quality index with historical monitored data in a Mediterranean Andean basin. Ecological Indicators. 2024;**166**:112373

[12] Ghorbani MA, Aalami MT, Naghipour L. Use of artificial neural networks for electrical conductivity modeling in Asi River. Applied Water Science. 2017;7:1761-1772

[13] Abozari N, Hassanvand M, Salimi AH, Heddam S, Mohammadi HO, Noori A. Comparison performance of artificial neural network based method in estimation of electric conductivity in wet and dry periods: Case study of Gamasiab River, Iran. Journal of Applied Research in Water and Wastewater. 2019;**6**(2):88-94

[14] Melesse AM, Khosravi K, Tiefenbacher JP, Heddam S, Kim S, Mosavi A, et al. River water salinity prediction using hybrid machine learning models. Water. 2020;**12**(10):2951

[15] Shah MI, Javed MF, Abunama T. Proposed formulation of surface water

quality and modelling using gene expression, machine learning, and regression techniques. Environmental Science and Pollution Research. 2021;**28**: 13202-13220

[16] Alqahtani A, Shah MI, Aldrees A, Javed MF. Comparative assessment of individual and ensemble machine learning models for efficient analysis of river water quality. Sustainability. 2022; **14**(3):1183

[17] Karbasi M, Ali M, Bateni SM, Jun C, Jamei M, Farooque AA, et al. Multi-step ahead forecasting of electrical conductivity in rivers by using a hybrid convolutional neural network-long short-term memory (CNN-LSTM) model enhanced by Boruta-XGBoost feature selection algorithm. Scientific Reports. 2024;**14**(1):15051

[18] Rajaee T, Khani S, Ravansalar M. Artificial intelligence-based single and hybrid models for prediction of water quality in rivers: A review. Chemometrics and Intelligent Laboratory Systems. 2020;**200**:103978

[19] Partal T, Küçük M. Long-term trend analysis using discrete wavelet components of annual precipitations measurements in Marmara region (Turkey). Physics and Chemistry of the Earth, Parts A/B/C. 2006;**31**(18): 1189-1200

[20] Zhou HC, Peng Y, Liang GH. The research of monthly discharge predictor-corrector model based on wavelet decomposition. Water Resources Management. 2008;**22**:217-227

[21] Penalba O, Rivera J, Bettolli ML. Trends and periodicities in the annual amount of dry days over Argentina, looking towards the climatic change. Options Méditerranéennes en ligne. 2010;**95**:27-33

[22] Nalley D. Analyzing trends in temperature, precipitation and streamflow data over Southern Ontario and Quebec using the discrete wavelet transform [thesis]. Department of Bioresource Engineering: McGill University: Montreal; 2012

[23] Steel EA, Lange IA. Using wavelet analysis to detect changes in water temperature regimes at multiple scales: Effects of multi-purpose dams in the Willamette River basin. River Research and Applications. 2007;**23**(4):351-359

[24] Ravansalar M, Rajaee T. Evaluation of wavelet performance via an ANN-based electrical conductivity prediction model. Environmental Monitoring and Assessment. 2015;**187**(6):366

[25] Khani S, Rajaee T. Modeling of dissolved oxygen concentration and its hysteresis behavior in rivers using wavelet transform-based hybrid models. CLEAN–Soil, Air, Water. 2017;**45**(2): 1-18

[26] Opsahl SP, Musgrove M. Occurrence of Pharmaceutical Compounds in the San Antonio Segment of the Edwards (Balcones Fault Zone) Aquifer, South-Central Texas, June 2018–August 2020. Reston, VA: U.S. Geological Survey; 2023

[27] Tokar AS, Johnson PA. Rainfall-runoff modeling using artificial neural networks. Journal of Hydrologic Engineering. 1999;**4**(3):232-239

[28] Masters T. Practical Neural Network Recipes in C++. San Francisco: Morgan Kaufmann; 1993

[29] Gutiérrez-Estrada JC, Vasconcelos R, Costa MJ. Estimating fish community diversity from environmental features in the Tagus estuary (Portugal): Multiple linear regression and artificial neural network approaches. Journal of Applied Ichthyology. 2008;**24**(2):150-162

[30] Haykin S. Neural Networks: A Comprehensive Foundation. Upper Saddle River, NJ: Prentice Hall PTR; 1998

[31] Hastie T, Tibshirani R, Friedman JH, Friedman JH. The Elements of Statistical Learning: Data Mining, Inference, and Prediction. New York: Springer; 2009

[32] Bishop CM, Nasrabadi NM. Pattern Recognition and Machine Learning. New York: springer; 2006

[33] Sazlı MH. A brief review of feed-forward neural networks. Communications Faculty of Sciences University of Ankara Series A2-A3 Physical Sciences and Engineering. 2006; **50**(01):11-17

[34] Rumelhart DE, Hinton GE, Williams RJ. Learning representations by back-propagating errors. Nature. 1986; **323**(6088):533-536

[35] Goodfellow I, Bengio Y, Courville A. Deep Learning. Cambridge, MA: MIT Press; 2016

[36] Chen Y, Song L, Liu Y, Yang L, Li D. A review of the artificial neural network models for water quality prediction. Applied Sciences. 2020;**10**(17):5776

[37] Hagan MT, Menhaj MB. Training feedforward networks with the Marquardt algorithm. IEEE transactions on Neural Networks. 1994;**5**(6):989-993

[38] Ranganathan A. The levenberg-marquardt algorithm. Tutoral on LM Algorithm. 2004;**11**(1):101-110

[39] Raghuwanshi NS, Singh R, Reddy LS. Runoff and sediment yield modeling using artificial neural networks: Upper Siwane River, India. Journal of Hydrologic Engineering. 2006;**11**(1):71-79

[40] Brion GM, Neelakantan TR, Lingireddy S. A neural-network-based classification scheme for sorting sources and ages of fecal contamination in water. Water Research. 2002;**36**(15):3765-3774

[41] Brion GM, Lingireddy S. Artificial neural network modelling: A summary of successful applications relative to microbial water quality. Water Science and Technology. 2003;**47**(3):235-240

[42] Lin GF, Wu MC, Chen GR, Tsai FY. An RBF-based model with an information processor for forecasting hourly reservoir inflow during typhoons. Hydrological Processes: An International Journal. 2009;**23**(25):3598-3609

[43] Ebrahimi H, Rajaee T. Simulation of groundwater level variations using wavelet combined with neural network, linear regression and support vector machine. Global and Planetary Change. 2017;**148**:181-191

[44] Adamowski J, Karapataki C. Comparison of multivariate regression and artificial neural networks for peak urban water-demand forecasting: Evaluation of different ANN learning algorithms. Journal of Hydrologic Engineering. 2010;**15**(10):729-743

[45] Mallat S. A Wavelet Tour of Signal Processing. San Diego: Academic Press; 1999

[46] Daubechies I. Ten Lectures on Wavelets. Philadelphia: Society for Industrial and Applied Mathematics; 1992

[47] Shensa MJ. The discrete wavelet transform: Wedding the a trous and Mallat algorithms. IEEE Transactions on Signal Processing. 1992;**40**(10): 2464-2482

[48] Kisi O. Wavelet regression model as an alternative to neural networks for

river stage forecasting. Water Resources Management. 2011;**25**:579-600

[49] Kişi Ö. Wavelet regression model as an alternative to neural networks for monthly streamflow forecasting. Hydrological Processes: An International Journal. 2009;**23**(25):3583-3597

[50] Rajaee T, Khani S. Comment on "performance of ANFIS versus MLP-NN dissolved oxygen prediction models in water quality monitoring a. Najah & a. El-Shafie & OA Karim & Amr H. El-Shafie. Environ Sci Pollut res (2014) 21: 1658-1670". Environmental Science and Pollution Research. 2016;**23**(1):938-940

[51] Rajaee T. Wavelet and ANN combination model for prediction of daily suspended sediment load in rivers. Science of the Total Environment. 2011; **409**(15):2917-2928

[52] Nourani V, Rahimi AY, Nejad FH. Conjunction of ANN and threshold based wavelet de-noising approach for forecasting suspended sediment load. International Journal of Information Technology. 2013;**3**:9-26

Chapter 3

Environmental Monitoring of Water and Lakes On-Site in Real-Time Using eDNA

Lars Eric Roseng, Nivedhitha Jothinarayanan, Leila Tajedin, Chau Ha Pham and Frank Karlsen

Abstract

Molecular environmental monitoring is essential for maintaining the health and sustainability of ecosystems, especially aquatic environments such as rivers, lakes, and oceans. Regular assessment of the molecular activities in these ecosystems is essential to identify changes in biodiversity caused by climate change, human activities, and the invasion of alien species. These factors can have profound effects on both the environment and human well-being. Traditional methods of environmental monitoring often involve manual sampling and laboratory analysis, which can be time-consuming, costly, and limited in scope. Recent advances in technology have led to the development of more efficient, automatic, real-time biomonitoring systems based on molecular activity, such as environmental DNA (eDNA) or RNA (eRNA) analysis. This chapter focuses on the status of new sampling methods, molecular techniques, microfluidic platforms, and Lab-on-a-Chip (LOC) technologies and provides a road-map for future efforts in automatic environmental monitoring systems.

Keywords: eDNA, environmental monitoring, invasive species, PCR, LAMP, automatic system

1. Introduction

The global temperature is increasing, weather patterns are becoming more unpredictable, natural habitats are disappearing, extreme weather events are occurring more frequently, and species diversity is under threat. These are the consequences of society's overconsumption of natural resources. Several major global reports from the UN Environmental Program (UNEP) confirm the drastic climate change and its significant impact on the environment and biodiversity, indicating that development is heading in the wrong direction [1–4]. Environmental sustainability means that we take care of the climate, nature, and other environmental values for future generations.

New technologies and methods for environmental monitoring can provide great leaps in terms of data collection, amount of information, and knowledge production. Better tools for mapping and monitoring will also provide data that can form the basis of more accurate research. The new tools must be able to monitor the state of the environment by delivering real-time molecular data on-site so that unwanted serious changes in the environment are detected early. It is important to establish more effective monitoring to better understand how human activities affect the environment, and what effects this influence has. The detection of the activity of the organisms that are typically sensitive to several of these environmental changes is also important to include.

In marine ecosystems, there is a need to further develop the understanding of what is due to influence from activities in the sea areas or adjacent coastal and land areas, for example, what is due to abnormal or human-created climate changes, and what is due to natural processes and variations. Global degradation of marine ecosystems or natural biological environments, composed with growing pressure to practice them, has formed a mandate for new, efficient, automatic, and cost-effective monitoring tools that make it possible to assess changes in the environmental state of freshwater, rivers, and the sea [5–7].

In the period 2019–2022, our research group at the University of South-Eastern Norway participated in a consortium where a system for automatic environmental monitoring in rivers and lakes using eDNA was to be developed. In this pre-commercial procurement project, the needs and requirements for the product solution had been specified by the Norwegian environmental authorities. The purpose of this chapter is to report the status and maturity of the technical solution that our consortium entered as a quotation in this supplier competition.

2. Control of harmful alien organisms

Humans have long played a role in moving various organisms to regions where they do not naturally belong. This intentional movement of organisms has been crucial to the development of agriculture and other industries, such as forestry, fishing, and horticulture. But today, due to global warming and new human activities, alien plants, animals, and other organisms pose one of the most important threats to biodiversity. The invasive species have contributed to more than half of the documented extinctions of animal species [8, 9].

Harmful alien organisms also incur significant economic costs. The global economic impact of invasive alien species, including crop, pasture, and forest loss as well as environmental destruction and control costs, exceeded $423 billion annually in 2019. This cost has at least quadrupled every decade since 1970 [10]. In Norway, the total annual economic costs of harmful alien species are estimated to be between 0.2 and 0.4 billion Euros [11]. It is important to note that some of the species that are conscious or have spread into the wild also have positive economic effects, such as those linked to better harvesting.

There are no signs that the number of new introductions of alien species is decreasing globally [12]. In Norway, as of today, a total of over 2700 alien species have been registered in the wild. The Norwegian Species Data Bank assesses that 1332 species pose an ecological risk and that 318 of these species pose a high or very high ecological risk [13]. An additional factor, which will affect the spread of alien harmful species in the future, is climate change, among other things. In Norway, increased temperature can improve the conditions for many alien species and make it easier for them to establish themselves and spread [14].

Global Biodiversity Outlook 5 (GBO5) [12] points out that while some progress has been made in addressing the challenges posed by harmful alien organisms, the progress observed in many countries remains insufficient. Particularly, efforts to prioritize and manage the import and spread routes for alien organisms have been notably weak. GBO5 identifies several measures that are particularly effective for the countries striving to meet their set goals. These measures include the dissemination of knowledge to various target groups about the impacts of alien organisms and the necessity of countermeasures, the development and availability of comprehensive risk assessments, and the intensification of efforts linked to the control of import routes. In addition, the establishment of a system for early detection and rapid response to new introductions and the creation and implementation of plans to combat already established harmful alien organisms are highlighted as crucial steps.

The Norwegian action plan for biological diversity [15] states that, in recent decades, there has been a significant reduction in the global natural diversity that affects the world's ecosystems. Some ecosystems are under extensive stress that they no longer provide the goods that people rely on. Although the situation for biological diversity in Norway, compared to other countries, is more positive in many areas, the action plan nevertheless identifies several areas of concern. It underlines the need to strengthen efforts against harmful alien organisms and to prioritize these efforts more strictly. The action plan emphasizes that comprehensive analyses must underpin the selection of measures, and that early, specific interventions should be prioritized. Furthermore, stimulating and solution-oriented measures should be put in place, and sectoral responsibilities for all national actors must be specified and clarified. The plan also emphasizes the importance of strengthening and further developing cooperation on measures against harmful alien organisms.

3. Automatic environmental monitoring using eDNA

To improve and streamline the environmental monitoring of water bodies, the Norwegian Environment Agency initiated a pre-commercial procurement project. This project aimed to exploit technological innovations for early molecular detection to carry out on-site real-time environmental monitoring, focusing on the potential of environmental DNA (eDNA) and environmental RNA (eRNA) technologies. The Norwegian Environment Agency allocates around 40 million Euros annually for monitoring and mapping Norwegian nature. Currently, environmental monitoring relies heavily on manual data collection by field workers at selected locations. Today's method is not only time-consuming and expensive but also provides limited opportunities to detect and prevent ecological damage and changes effectively. Therefore, there is a significant need to improve and automate environmental monitoring processes.

Recent advancements in detecting species-specific eDNA and eRNA have shown promise in improving environmental monitoring. These methods simplify and enhance sample collection, allowing for the identification and cataloging of a broader range of species within a water sample. Moreover, eRNA analysis helps map the gene activities, providing insights into the ecological impacts of external factors [16–18]. Aquatic environments are particularly well suited for these techniques.

One of the main challenges, but at the same time the potential, when using eDNA for environmental monitoring is to establish automatic measurements. There is both a national and international need and a market for technology that meets the need for early molecular detection, warning, and effective environmental monitoring. This approach can potentially offer early detection of harmful organisms, such as invasive

Environmental DNA (eDNA) and environmental RNA (eRNA) offer new techniques for species detection that involve genetic analysis of environmental samples. This method enables the identification of species without their visual sighting by means of collecting samples from water, air, silt, or ground. The genetic material access free of charge is ample and has been categorized as Intracellular, intraorganellar, dissolved, or particle encapsulated DNA and RNA. These can be produced by different sources such as by skin, mucus, saliva, secretion, gametes, excretions, blood, plant materials, dead organisms, and even complete microbes.

Figure 1.
Facts on environmental DNA/eRNA [19–26].

species, parasites, and pathogens, enabling timely interventions to mitigate severe ecological consequences (**Figure 1**).

Environmental DNA (eDNA) and environmental RNA (eRNA) offer new techniques for species detection that involve genetic analysis of environmental samples. This method enables the identification of species without their visual sighting by means of collecting samples from water, air, silt, or ground. The genetic material access free of charge is ample and has been categorized as Intracellular, intraorganellar, dissolved, or particle-encapsulated DNA and RNA. These can be produced by different sources such as by skin, mucus, saliva, secretion, gametes, excretions, blood, plant materials, dead organisms, and even complete microbes.

3.1 The pre-commercial procurement competition

The purpose of a pre-commercial procurement project is to promote demand-based innovation adapted to public sector challenges and to exploit these challenges for value creation in the business community. This involves cooperation between the public and private sectors to meet the actual requirements and achieve mutual benefit. The procurement process itself follows the pre-commercial procurement phase.

The pre-commercial procurement competition, called DNA Auto, was initiated in 2019 when the Norwegian Environment Agency invited relevant stakeholders to a dialog conference to identify the optimal solution and clarify potential and costs associated with automatic environmental monitoring [27].

3.2 The customer's needs and requirements

Participants in the DNA Auto competition were tasked with developing a product solution that met the specific needs and requirements outlined by the environmental authorities [28, 29]. This included providing comprehensive information on technical solutions currently available in the market, the potential and limitations of these solutions, challenges in developing automatic environmental monitoring systems, estimated development and final product costs, development timelines, and potential areas of application of the developed solutions.

The envisaged product system had to work under various freshwater conditions, targeting specific organisms relevant to environmental management. Key requirements included tools to monitor the spread of harmful alien species [8, 30] and disease organisms such as *Gyrodactylus salaris* [29], Batrachochytrium dendrobatidis [31], and

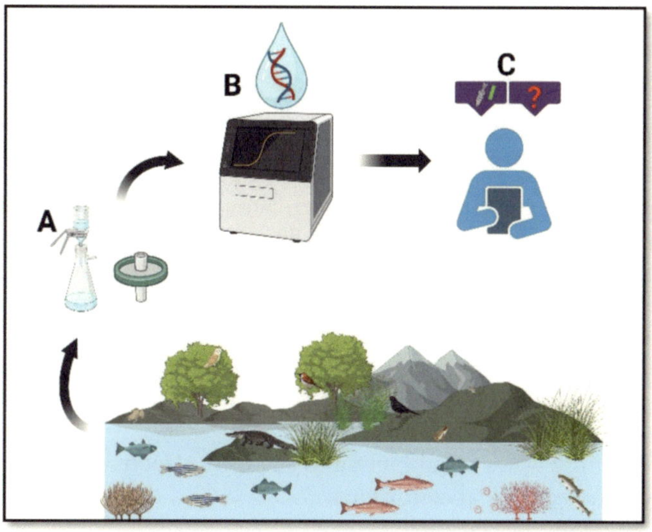

Figure 2.
Illustration of the three functions in a comprehensive product system for automatic environmental monitoring. A. Sampling/filtration, B. eDNA/eRNA detection, C. Transfer of digital data.

crayfish plague [32]. Furthermore, the system was expected to monitor endangered species and to classify ecological conditions according to the EU's Water Framework Directive [33].

Three primary sub-goals were established for the competition:

- Develop an automatic sampling system that satisfies the requirements for standardization of sample material.
- Develop an automatic analysis unit capable of monitoring eDNA (and/or eRNA) and connectable to the instinctive sample collector.
- Create an automatic transfer and processing of data for the generation of alerts regarding hazardous biological agents.

The tender documents contained a description of the customer's needs and requirements that were to be met when the new technology for automatic environmental monitoring was to be developed.

In addition, it was desirable that each automated step/part could be used independently, whether the other steps were automated or not. However, there had to be a plan for how all steps could be used together to create a comprehensive system for environmental monitoring using eDNA. In this way, each part that is automated will gradually be able to contribute to improving and making environmental monitoring more efficient (**Figure 2**).

3.3 The advantages of automatic equipment

Automation of environmental monitoring will be important for society, the environment, and private businesses. The establishment and spread of disease-causing organisms or foreign harmful species already entail significant socio-economic costs, and early intervention is essential. National and international regulations and obligations also lay down guidelines for the implementation of environmental monitoring. Methodology and technology for early detection, warning, and modeling of the spread of alien organisms are therefore also priority research areas. The eDNA Auto project will therefore have great potential for value creation in the business world, for significant socio-economic impact, and for achieving better public services.

In the eDNA Auto project, there was initially a desire to focus on a tool to monitor the spread of harmful alien organisms, especially pike, and the parasite *Gyrodactylus salaris* in Norwegian waterways. Simultaneously, there was an aim to confirm that the tool could easily be extended to other groups of organisms and problems.

Other relevant areas of use for automated eDNA sample collection and analysis may be:

- Monitoring of the water regulations and ensuring effective monitoring of waterways, and ecological conditions.
- Early detection of pollution from petroleum operations and other future offshore industries.
- Monitoring and intervention/treatment in terrestrial systems.
- Monitoring the spread of human pathogenic organisms and antimicrobial resistance.

Automatic monitoring and intervention tools would expand the likelihood of early-stage detection of potential hazards. Automatic sample collection will also contribute to better agreeable and reproducible sample collection.

4. Automatic eDNA sampling and filtering

4.1 Need for standardization when eDNA is collected

Currently, there are no standard guidelines for eDNA technology for environmental monitoring. Many different manual collection methods for eDNA in water samples have been developed, and the methodology differs from laboratory to laboratory. The water samples can either be brought to the laboratory for filtration, or they can be filtered, separated, concentrated, or sorted directly on-site. The filtration volumes vary from 15 to 6 L and are even higher for eDNA (or eRNA) collection [34, 35]. Filtration of larger volumes is often challenging, due to filter clogging from various particles in river or lake water. However, several new technologies including smart treatment of the water will increase the possibility to detect all important targets.

If an automatic sampling and filtering method is to be established on-site, the collection method should first be standardized. Quality assurance practices necessitate the need for standard operating procedures (SOPs) to minimize bias and errors. Numerous eDNA methods identified in a 2018 meta-analysis were classified as non-reproducible, primarily due to deficiencies in accurate and complete reporting, or because subjective protocols were used, only 5% of the studies assessed were replicable [32]. Although the collection and detection of eDNA was a developing discipline at this time, such a low reproducibility rate is of concern for academic research.

Standardized methods must be safe, cost-effective, simple, able to last over time, yet adaptable, and based on available data. Given the myriads of different approaches, tools, and procedures used to collect and filter eDNA from water, it is possible that some methods in use have a greater impact on data quality than others. It should be noted that an optimized method has not yet been developed, which can be used for all possible scenarios due to differences in species behavior, different habitat conditions, and requirements for sufficient reproducibility [32]. However, standardization of species-specific methods for different purposes can be achieved if this is combined into reflective SOPs.

Another fundamental aspect of eDNA or eRNA collection is the sampling strategy. Statistical sampling strategies for river systems have been designed for eDNA biomonitoring, guided by the spatial distribution of species and the availability of sampling sites [33]. The number and location of sampling sites should be carefully considered, and it has been suggested that eDNA sampling should be scaled to match the size of the watercourse to provide reliable estimates [36]. The recommendations also indicate that hydrological conditions in the water bodies can affect eDNA dispersion, and that targeted sampling can result in an increased probability of positive detections for a given target species.

eDNA can be found in a range of particle sizes [37] depending on its biological source, origin, and degradation. Choice of filter size has been shown to affect the amount of eDNA captured and subsequent sensitivity [38], for example, small pore size filters can easily become clogged in turbid water by suspended material which can introduce high levels of PCR inhibition [39]. Inhibition in freshwater can also cause non-amplification of high eDNA copy numbers [40], although careful PCR reagent selection can reduce the effect of inhibitors such as humic, fulvic, and tannic acids [41]. Different filter materials have also been shown to yield varying levels of eDNA due to different DNA binding affinities [42] but both glass fiber [43], cellulose-based [44], and polyethersulfone [45] filters have been shown to be useful in various studies.

When choosing a standard method for eDNA collection and filtering, it is also important to be aware of the advantages and disadvantages of different filter sizes. It has even been suggested that optimal methods may not exist for eDNA collection due to the variability introduced by the many possible target species and environmental factors [35].

4.2 Lysis of the eDNA on the filter

In connection with the filtration, sediments and other solid particles in the water will accumulate on the filter surface. It is among these particles that the eDNA or eRNA material ends up, and much of this eDNA or eRNA is probably particle-bound. In further sample preparation, the next steps are eDNA and/or eRNA purification and extraction. There are five basic steps in the DNA extraction process that are consistent across all possible DNA purification chemistries:

1. Disruption of the cell structure to create a lysate.

2. Separation of the soluble DNA from cell debris and other insoluble materials.

3. Binding of DNA of interest to a purification matrix or micro-nano beads.

4. Wash proteins and other contaminants away from the matrix.

5. Elution of DNA.

When developing an automatic product system, lysis must take place directly on the filter or in relation to direct bead-based binding. After filtering the required amount of water, the inlet channel for the water in the filter unit must be closed, and residual water drained, before a lysis buffer is added to the filter. This process can be carried out in a closed filter unit with two inlets.

The lysis of the main filter (the basic step 1) must be followed by the basic steps 2–5, including washing and DNA elution, as part of the complete sample analysis. The purpose of lysis is to disrupt/break up cells, mitochondria, or other membrane-bound biological particles trapped on the filter surface quickly and completely to release the nucleic acids (DNA/RNA). When the biological membranes are broken up, the chemical content of these biological particles is made available for further analysis [46, 47].

After lysis of the biological particles on the filter, the lysate (which will contain the eDNA or eRNA) must be transferred to a separate reaction chamber containing the purification matrix or beads. The lysate in the filter unit (product part A) is then transferred to the analysis instrument (product part B) to carry out the eDNA analysis.

Common cell lysis methods include mechanical disruption, liquid homogenization, high-frequency sound waves, and reagent-based (chemical) methods. In many protocols, a combination of chemical disruption and another method is often used since chemical disruption of cells rapidly inactivates proteins, including nucleases.

4.3 Complexity of water in the ecosystem

Water bodies such as rivers and lakes are dynamic ecosystems that show considerable complexity in their physical, chemical, and biological properties. The complexity arises from the interaction between natural processes and human activities that affect water quality, ecosystem health, and hydrological behavior [48], and the content and quantity of particles vary with the season. Rivers and lakes are drinking water for many living organisms, and at the same time are habitats and provide nutrients for other species.

Factors such as nutrient levels, pollutants, biodiversity, hydrological patterns, and changes in land use provide an understanding of the complexity [49]. Rivers and lakes are also interconnected systems where the water is continuously moving. These processes include water exchange, sediment transport, and energy transfer. The rivers transport water from high to low areas and shape the landscape, which affects the water quality in the lake.

4.4 Importance of pre-filtering

A pre-filtration step is important in a fluid system to preserve and protect downstream filters and components from damage and premature failure, thereby improving system life and reliability. A suitable continuously flowing pre-filtration solution must therefore be selected and dimensioned to increase the system's filtration efficiency.

The first pre-filter should capture larger debris and allow other downstream pre-filters to be selected to remove a specific range of particle sizes, allowing each stage of filtration to do its part as designed. In a continuously flowing pre-filtration system, the largest and unwanted particles must be separated out, so that the eDNA from a specified amount of water—preferably a volume of 5 liters—can flow through the main filter without clogging. Our research group has previously developed a continuously flowing countercurrent technology that can separate and concentrate particles from complex liquids in specific sizes from 25 μm and upward [50, 51]. Such continuously flowing microfluidic pre-filters can be assembled in series to remove different particles based on their sizes.

The first pre-filter should capture larger debris and allow downstream continuous flowing pre-filters to be selected to remove a specific range of smaller particle sizes, allowing each stage of filtration to do its part as designed. In a continuously flowing pre-filtration system, the largest and unwanted particles must be removed, so that the eDNA from a specified amount of water—preferably a volume of 5 liters—can flow through the main filter without clogging. Our research group has previously developed a continuously flowing countercurrent technology that can separate and concentrate particles from complex liquids in specific sizes from 25 μm and upward [50, 51]. Such continuously flowing microfluidic pre-filters can be assembled in series to remove different particles based on their sizes.

The first pre-filter should capture larger debris and allow other downstream pre-filters to be selected to remove a specific range of particles so that each stage of pre-filtration does its part as designed. In a continuously flowing pre-filtration system, the largest and unwanted particles are removed, so that the eDNA from a specified amount of water—preferably a volume of 5 liters—can flow through the main filter without clogging. Previously, our research group has developed a continuously flowing countercurrent technology that can separate and concentrate particles from complex liquids in specific sizes from 25 μm and upward [50, 51]. Such continuously flowing microfluidic pre-filters can be assembled in series to remove different particles based on their sizes.

A pre-filtration system based on this counter-flow technology has not yet been fully developed and tested, but in theory, this technology is promising. Unwanted particles can be removed in a continuously flowing process and ensure that the main filter is not blocked by clogging particles. Unwanted particles can be refined out of the complex water liquid using this counter-flow technology.

To develop such a pre-filtering technology, more research is needed to better understand what eDNA and eRNA are. Is eDNA or eRNA available as free DNA or RNA dissolved in the water, or are they bound to specific particles in the water, and if so, which particles are these? Here we lack knowledge today. To develop a continuously flowing automatic pre-filtration unit for obtaining eDNA in complex liquids, it is necessary to standardize this entire filtering process.

5. Automatic analysis of eDNA

Today's methods for environmental monitoring are mainly based on manual registrations by field workers who cover small selections of localities in nature. Analysis of the samples is then carried out by specialist laboratories. This is a time-consuming, expensive, and slow manual process that gives limited opportunity to detect and follow changes in the environment quickly enough.

5.1 Lab-on-a-chip technology

The development of Lab-on-a-chip (LOC) technologies has been closely linked to developments in the field of microfluidics. Microfluidics relates to the study of the behavior of fluids through microchannels and to the fabrication of miniaturized devices containing chambers and channels through which fluids flow.

The ultimate goal of LOC technologies is to integrate one or more laboratory processes or techniques on a single miniaturized chip [52–54]. It is theoretically possible to accomplish nearly any laboratory task on a smaller scale with LOC technology. This can include chemical synthesis, clinical diagnostics, biomarker validation, DNA sequencing, and biochemical detection.

Compared to conventional laboratory benchtop operations, lab-on-a-chip technology has a number of potential benefits. In medical offices or other remote sites with limited access to laboratory facilities and sophisticated equipment, a portable, handheld device that can perform point-of-care clinical diagnostics could be useful.

Due to lower reagent usage and chemical waste, standard benchtop operations can now be miniaturized with benefits for both the environment and financial effectiveness. The integration of microchannels can theoretically achieve great parallelization, allowing multiple analyses to occur concurrently on a single device.

Point-of-need (PoN) or point-of-care (PoC) are technology systems directly related to real-time on-site operations [55, 56]. The necessity of switching from traditional lab-centric diagnostics to point-of-care (PoC) settings has been highlighted by the recent coronavirus (COVID-19) epidemic. Modern LOC technologies make it easier to translate into point-of-care (PoC) environments by automating, integrating, and shrinking several test functions onto a single disposable chip.

Extensive research is now being done on both paper-based analyzers and microfluidic platforms. A lot of attention is being paid to enhancing the multiplexing and automation capabilities of the platforms while also making their design simpler.

5.2 The choice of DNA amplification method

Initially, the detection and analysis of eDNA required sophisticated laboratory equipment and specialized skills. Early methods focused on Polymerase Chain Reaction (PCR) to amplify DNA sequences for identification. While effective, PCR-based methods have limitations, particularly in field settings where laboratory infrastructure may be lacking. This has led to the development of alternative techniques, such as Loop-mediated Isothermal Amplification (LAMP), which offer faster and more accessible options for eDNA analysis.

At the moment, PCR is the most popular and versatile DNA identification method for keeping an eye on the aquatic environment [57, 58]. The inadequacy of PCR depends on the fact that it necessitates sophisticated laboratory equipment and technical know-how, both of which are frequently lacking in underfunded facilities. Prior research has thoroughly examined the benefits and drawbacks of both PCR and LAMP [59]. The drawbacks of PCR-based approaches can be effectively addressed by the relatively new and distinctive LAMP technology.

LAMP techniques are quick and easy ways to overcome the potential for complicated biological samples to contain inhibitory substances that could interfere with the amplification process [60–63]. LAMP may convert double-stranded DNA into single-stranded DNA without the need for a heat denaturation step, in contrast to PCR. The complementary template is created by DNA polymerase activity with significant displacement. It is therefore technologically easier to use an isothermal amplification process such as LAMP inside the LOC platform, then the temperature is kept constant (65°C) during the whole amplification process. A disadvantage of LAMP is that this method is relatively new. LAMP has therefore been far less used compared to PCR. The established gene libraries for verified PCR primers are therefore well developed, while verified LAMP primers are almost not available. Another notable characteristic of the LAMP technique is good specificity and high amplification efficiency.

5.3 Development of the Fordetect technology

The Fordetect system has been developed by a research group at the department of Microsystems of USN, led by Professor Frank Karlsen. The product system is a lab-on-a-chip-based technology that enables real-time automatic sample preparation and purification of DNA/RNA in combination with automatic amplification and detection [64–68]. The development of this enabling technology started in 2002, and many institutes and companies from various European countries have been involved. The aim has been to deliver a product solution to make DNA and RNA analyses available everywhere (on-site, real-time analysis), to ensure better health services and better environmental monitoring systems.

The project is the result of many years of ongoing research to improve the analysis of biological samples. Some of the later results were a new prototype of the entire analysis system (1) a new LOC platform, (2) an integrated microfluidic system in a disposable cassette, and (3) a newly developed electronic instrument (**Figure 3**). This product system is currently at technology readiness level (TRL) 5.

The instrument has software that manages and controls all the biochemical processes inside the developed automatic microfluidics system. The instrument's software closes and opens valves in the correct order, pushes liquid back and forth, collects washing buffers, sends used chemicals to waste, elutes the eDNA in the sample, runs the amplification process, records the analysis result, and forwards this to a mobile phone. A patent application for this entire microfluidic system was filed in March 2021 [64–68].

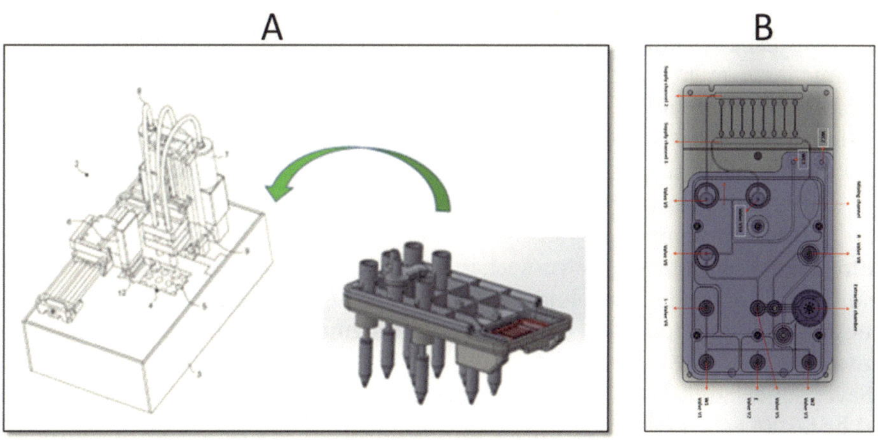

Figure 3.
Existing prototypes of today's Fordetect product system that combine purification of total DNA/RNA and real-time isothermal amplification and detection. A. The Fordetect system and the disposable cassette. B. The lab-on-a-chip part of the disposable cassettes.

Testing of this Fordetect system has shown promising results using different biomarkers. All required LAMP reagents are ready inside the LOC unit as lyophilized reagents. This real-time product system has been successfully demonstrated. There is still a lot of testing and further product development that remains to bring this automatic DNA analysis system through the final TRL step toward the market.

The research made it possible to build microchannels, microvalves, and micropumps and construct the first functional microfluidic-based LOC chips. One challenge was to get the biologists and medical researchers to talk to physicists and those with knowledge of electronics. The interdisciplinary environment that could do this was found at the Institute of Microtechnology in Mainz (IMM), and the first microfluidic-based system that could extract or purify total RNA and DNA was presented. It was now possible to explore microfluidic amplification and detection systems that combined purification of total DNA/RNA and real-time isothermal amplification and detection.

6. Conclusion

Ecological health and sustainability are important for global well-being and future generations. Global biodiversity is today under threat, and there are many indications that the sixth mass extinction is developing [69, 70]. The global environmental footprint of humans has become too large, and irreversible changes are occurring in the ecosystems. It has become a fact that conventional manual monitoring methods have limitations related to time frame, costs, and efficiency. Without new technology that automatically monitors the ecosystems in real time, it is not possible to respond quickly enough concerning the ongoing changes taking place in nature.

Environmental DNA (eDNA or eRNA) analysis has emerged as a powerful tool for monitoring changes in ecosystems. Implementation of cost-effective on-site monitoring systems that can deliver real-time molecular environmental data is necessary to assess the complexity of unwanted alien species and pathogens. The concept of automatic early detection followed by rapid intervention and treatment is essential in preventive species management and in monitoring the health of ecosystems.

The environmental authorities' needs and requirements for a product system for automatic environmental monitoring using eDNA in rivers and lakes will consist of three parts: (1) a system for automatic sampling/filtering, (2) an analysis unit (an electronic instrument), and (3) an automatic unit for transfer and processing of electronic environmental data.

When it comes to the technological state-of-the-art and the maturity of a holistic product solution for this automatic technology, our work shows that there is a need for both more basic and applied research. Basic research is important to standardize a good method for sampling/filtering eDNA. The microfluidic LOC technology is mature and many suitable electromechanical components and software are available. What remains in product development is more testing and further development of the actual interaction between the electromechanical components and the LOC technology.

Several mechanical components and critical functions in the analysis instrument must be tested and further developed for better control and operation of the molecular biological processes inside the LOC units. The important thing is to get valves, actuators, pumps, and the microfluidic system, including chemicals and reagents, inside the LOC platform to work perfectly together. More testing and further product development is required to bring the prototypes through the TRL levels.

DOI: http://dx.doi.org/10.5772/intechopen.1007780

Acknowledgements

We acknowledge everyone who has been involved in the development of this automatic environmental monitoring system.

Conflict of interest

All authors declare no conflict of interest.

Author details

Lars Eric Roseng*, Nivedhitha Jothinarayanan, Leila Tajedin, Chau Ha Pham and Frank Karlsen
Department of Microsystems, University of South-Eastern Norway, Raveien, Norway

*Address all correspondence to: lars.roseng@usn.no

References

[1] Global-Resource-Outlook [Online]. 2024. Available from: https://wedocs.unep.org/20.500.11822/44901

[2] Beyond an Age of Waste Turning Rubbish into a Resource [Online]. 2024. Available from: https://wedocs.unep.org/20.500.11822/44939

[3] UNEP Annual Report [Online]. Available from: https://www.unep.org/annualreport/2023

[4] Global Waste Management Outlook 2024. UNEP - UN Environment Programme. Available from: https://www.unep.org/resources/global-waste-management-outlook-2024

[5] Hyvärinen H et al. Cost-efficiency assessments of marine monitoring methods lack rigor—A systematic mapping of literature and an end-user view on optimal cost-efficiency analysis. Environmental Monitoring and Assessment. 2021;**193**:400. DOI: 10.1007/s10661-021-09159-y

[6] The European Maritime Spatial Planning Platform. DEVelopment of Innovative Tools for Understanding Marine Biodiversity and Assessing Good Environmental Status. Available from: https://maritime-spatial-planning.ec.europa.eu/projects/development-innovative-tools-understanding-marine-biodiversity-and-assessing-good

[7] Azti A-B et al. Innovative and practical tools for monitoring and assessing biodiversity status and impacts of multiple human pressures in marine systems. Environmental Monitoring and Assessment. 2023;**196**(8):694

[8] Forslag til tiltaksplan for bekjempelse av skadelige fremmede organismer. 2019

[9] Home. Convention on Biological Diversity [Online]. Available from: https://www.cbd.int/

[10] Invasive Alien Species Report. UNEP - UN Environment Programme [Online]. Available from: https://www.unep.org/resources/report/invasive-alien-species-report

[11] Samfunnsøkonomiske kostnader ved fremmede arter i Norge: Metodeutvikling og noen foreløpige tall [Online]. Available from: https://www.vista-analyse.no/no/publikasjoner/samfunnsokonomiske-kostnader-ved-fremmede-arter-i-norge-metodeutvikling-og-noen-forelopige-tall/

[12] Global Biodiversity Outlook 5. Convention on Biological Diversity [Online]. Available from: https://www.cbd.int/gbo5

[13] Alien Species List 2023 - Species Data Bank [Online]. Available from: https://artsdatabanken.no/lister/fremmedartslista/2023?TaxonRank=AssessedAtSameRank

[14] Forsgren E et al. Klimaendringenes påvirkning på naturmangfoldet i Norge. NINA Report 1210. Trondheim, Norway; 2015. 133 p. Available from: https://www.miljodirektoratet.no/globalassets/publikasjoner/m443/m443.pdf. ISBN: 978-82-426-2840-4

[15] Natur for livet Norsk handlingsplan for naturmangfold [Online]. Available from: www.fagbokforlaget.no/offpub

[16] Littlefair J-E et al. Environmental nucleic acids: A field-based comparison for monitoring freshwater habitats using eDNA and eRNA. Molecular Ecology Resources. 2022;**22**(8):2928-2940

[17] Miyata K et al. Comparative environmental RNA and DNA metabarcoding analysis of river algae and arthropods for ecological surveys and water quality assessment. Scientific Reports. 2022;**12**(1):19828

[18] Veilleux H-D et al. Environmental DNA and environmental RNA: Current and prospective applications for biological monitoring. Science of the Total Environment. 2021;**782**:146891

[19] Taberlet P et al. Environmental DNA. Molecular Ecology. 2012;**21**(8):1789-1793

[20] Ficetola G-F et al. Species detection using environmental DNA from water samples. Biology Letters. 2008;**4**(4):423-425

[21] Thomsen P-F et al. Detection of a diverse marine fish Fauna using environmental DNA from seawater samples. PLoS One. 2012;**7**(8):41732

[22] Clare E-L et al. eDNAir: Proof of concept that animal DNA can be collected from air sampling. PeerJ. 2021;**9**:e11030

[23] Jackson M et al. Using nested PCR to improve detection of earthworm eDNA in Canada. Soil Biology and Biochemistry. 2017;**113**:215-218

[24] Yasashimoto T et al. Environmental DNA detection of an invasive ant species (*Linepithema humile*) from soil samples. Scientific Reports. 2021;**11**(1):1-12

[25] Mauvisseau Q et al. The multiple states of environmental DNA and what is known about their persistence in aquatic environments. Environmental Science & Technology. 2022;**2022**:5333

[26] Clare E-L et al. Measuring biodiversity from DNA in the air. Current Biology. 2022;**32**(3):693-700

[27] Bohmann K et al. Environmental DNA for wildlife biology and biodiversity monitoring. Trends in Ecology & Evolution. 2014;**29**(6):358-367

[28] Barnes M-A et al. The ecology of environmental DNA and implications for conservation genetics. Conservation Genetics. 2015;**17**(1):1-17

[29] Invitation to Market-Related Dialogue [Online]. Available from: https://www.miljodirektoratet.no/aktuelt/arrangementer/dialogkonferanse---

[30] Automatic Environmental Monitoring Using Environmental-DNA Appendix B Customer Needs and Requirements for the Idea Sketch [Online]. Available from: http://tema.miljodirektoratet.no/Documents/Nyhetsdokumenter/M-2882014_rapport_nett.pdf

[31] Fisher M-C et al. Global emergence of *Batrachochytrium dendrobatidis* and amphibian chytridiomycosis in space, time, and host. Annual Review of Microbiology. 2009;**63**:291-310

[32] Crayfish Plague (*Aphanomyces astaci*) [Online]. Available from: https://www.vetinst.no/en/surveillance-programmes/crayfish-plague-aphanomyces-astaci

[33] Water Framework Directive - European Commission [Online]. Available from: https://environment.ec.europa.eu/topics/water/water-framework-directive_en

[34] Wilcox T-M et al. Understanding environmental DNA detection probabilities: A case study using a stream-dwelling char *Salvelinus fontinalis*. Biological Conservation. 2016;**194**:209-216

[35] Hinlo R et al. Methods to maximise recovery of environmental DNA

from water samples. PLoS One. 2017;**12**(6):e0179251

[36] Lewis M et al. The forensic potential of environmental DNA (eDNA) in freshwater wildlife crime investigations: From research to application. Science & Justice. 2024;**64**(4):443-454

[37] Carraro L et al. How to design optimal eDNA sampling strategies for biomonitoring in river networks. Environmental DNA. 2021;**3**(1):157-172

[38] Altermatt F et al. Quantifying biodiversity using eDNA from water bodies: General principles and recommendations for sampling designs. Environmental DNA. 2023;**5**(4):671-682

[39] Turner C-R et al. Particle size distribution and optimal capture of aqueous macrobial eDNA. Methods in Ecology and Evolution. 2014;**5**(7):676-684

[40] Cooper M-K et al. Practical eDNA sampling methods inferred from particle size distribution and comparison of capture techniques for a critically endangered elasmobranch. Environmental DNA. 2022;**4**(5):1011-1023

[41] Kumar G et al. One size does not fit all: Tuning eDNA protocols for high- and low-turbidity water sampling. Environmental DNA. 2022;**4**(1):167-180

[42] Jane S-F et al. Distance, flow and PCR inhibition: eDNA dynamics in two headwater streams. Molecular Ecology Resources. 2015;**15**(1):216-227

[43] Uchii K et al. Comparison of inhibition resistance among PCR reagents for detection and quantification of environmental DNA. Environmental DNA. 2019;**1**(4):359-367

[44] Liang Z et al. Filtration recovery of extracellular DNA from environmental water samples. Environmental Science & Technology. 2013;**47**(16):9324-9331

[45] Lacoursière-Roussel A et al. Estimating fish abundance and biomass from eDNA concentrations: Variability among capture methods and environmental conditions. Molecular Ecology Resources. 2016;**16**(6):1401-1414

[46] Islam M-S et al. A review on macroscale and microscale cell lysis methods. Micromachines. 2017;**8**(3):83

[47] Brown R-B et al. Current techniques for single-cell lysis. Journal of the Royal Society Interface. 2008;**5**(2):131-138. DOI: 10.1098/rsif.2008.0009.focus

[48] Biggs J et al. The importance of small waterbodies for biodiversity and ecosystem services: Implications for policy makers. Hydrobiologia. 2016;**793**(1):3-39

[49] Zhang Q et al. The influence of dam and lakes on the Yangtze River streamflow: Long-range correlation and complexity analyses. Hydrological Processes. 2012;**26**(3):436-444

[50] Tran-Minh N et al. Design and optimization of non-clogging counter-flow microconcentrator for enriching epidermoid cervical carcinoma cells. Biomedical Microdevices. 2011;**13**(1):179-190

[51] Dong T et al. Integratable non-clogging microconcentrator based on counter-flow principle for continuous enrichment of CaSki cells sample. Microfluidics and Nanofluidics. 2011;**10**(4):855-865

[52] Team Elveflow. Introduction to Lab-on-a-Chip 2023: Review, History and Future. Paris, France; 2023 [Online]. Available from: https://www.elveflow.com/microfluidic-reviews/general-microfluidics/introduction-to-lab-on-a-chip-review-history-and-future/

[53] Bahnemann J et al. Microfluidics in biotechnology: Overview and status quo. Advances in Biochemical Engineering/ Biotechnology. 2022;**179**:1-16

[54] Enders A et al. Towards small scale: Overview and applications of microfluidics in biotechnology. Molecular Biotechnology. 2024;**66**(3):365-377

[55] Choi S. Powering point-of-care diagnostic devices. Biotechnology Advances. 2016;**34**(3):321-330

[56] Hansen S et al. Point-of-care or point-of-need diagnostic tests: Time to change outbreak investigation and pathogen detection. Tropical Medicine and Infectious Disease. 2020;**5**(4):151

[57] Dejean T et al. Persistence of environmental DNA in freshwater ecosystems. PLoS One. 2011;**6**(8):8-11

[58] Antoinette MLA et al. Detecting an elusive invasive species: A diagnostic PCR to detect Burmese python in Florida waters and an assessment of persistence of environmental DNA. Molecular Ecology Resources. 2013;**14**(2):374-380

[59] Lau H-Y et al. Advanced DNA-based point-of-care diagnostic methods for plant diseases detection. Front Plant Sci. 2017;**8**:1-14

[60] Mori Y et al. Loop-mediated isothermal amplification (LAMP): Recent progress in research and development. Journal of Infection and Chemotherapy. 2013;**19**(3):404-411

[61] Niessen L. Current state and future perspectives of loop-mediated isothermal amplification (LAMP)-based diagnosis of filamentous fungi and yeasts. Applied Microbiology and Biotechnology. 2015;**99**:553-574. DOI: 10.1007/s00253-014-6196-3

[62] Foo P-C et al. Loop-mediated isothermal amplification (LAMP) reaction as viable PCR substitute for diagnostic applications: A comparative analysis study of LAMP, conventional PCR, nested PCR (nPCR) and real-time PCR (qPCR) based on *Entamoeba histolytica* DNA derived from. BMC Biotechnology. 2020;**20**(1):1-15

[63] Notomi T et al. Loop-mediated isothermal amplification of DNA. Nucleic Acids Research. 2000;**28**(12):e63. DOI: 10.1093/nar/28.12.e63

[64] Gulliksen A et al. Towards a sample-in, answer-out point-of-care platform for nucleic acid extraction and amplification: Using an HPV E6/ E7 mRNA model system. Journal of Oncology. 2012;**2012**:905024

[65] Gulliksen A et al. Real-time nucleic acid sequence-based amplification in Nanoliter volumes. Analytical Chemistry. 2004;**76**(1):9-14

[66] Baier T et al. Hands-free sample preparation platform for nucleic acid analysis. Lab on a Chip. 2009;**9**(23):3399-3405

[67] Gulliksen A et al. Microchips for the diagnosis of cervical cancer. Methods in Molecular Biology. 2007;**385**:65-86

[68] Gulliksen A et al. Parallel nanoliter detection of cancer markers using polymer microchips. Lab on a Chip. 2005;**5**(4):416-420

[69] Cowie R-H et al. The sixth mass extinction: Fact, fiction or speculation? Biological Reviews. 2022;**97**(2):640-663

[70] What is the Sixth Mass Extinction and what can we do about it? Stories. WWF. [Online]. Available from: https://www.worldwildlife.org/stories/what-is-the-sixth-mass-extinction-and-what-can-we-do-about-it

Section 3

Improved Modeling Approaches

Chapter 4

Soil Erosion Potential Model in Tropical Catchment

Mandana Abedini, Md Azlin Md Said and Fauziah Ahmad

Abstract

Soil erosion is a significant environmental challenge, especially in tropical regions where heavy rainfall and land use changes accelerate soil degradation. Effective land management and conservation strategies require understanding and predicting soil erosion potential. This study presents a Soil Erosion Potential Model (SEPM) tailored for tropical catchments, integrating remote sensing and Geographic Information Systems (GIS) with the Revised Universal Soil Loss Equation (RUSLE). The Ulu Kinta Catchment in Malaysia, spanning 30,752 hectares and linked to the Ulu Kinta Dam in Ipoh, serves as a case study. To estimate the Rainfall (R) factor, rainfall data was correlated with topographical variables using rain gauge data and the downscaled tropical rainfall measuring mission (TRMM) 3B43 dataset over 11 years. Fieldwork involved soil sampling to determine the Soil Erodibility (K) factor. Shuttle radar topography mission (SRTM) data assessed topographical effects through Digital Elevation Models (DEMs), and Normalized Different Vegetation Index (NDVI) alongside Land Use/Land Cover (LULC) data from SPOT5 imagery was used to estimate the Cover (C) and Support Practice (P) factors. The annual soil erosion map indicated a mean erosion rate of 34.72 tons per hectare per year, peaking at 150 tons per hectare per year, associated with steep slopes, high rainfall, and insufficient support practices. Alarmingly, 19.98% of the catchment experienced severe to extremely severe erosion rates. The SEPM model was validated against previous studies, achieving a validation rate of 71.9%, demonstrating a reasonable correlation with similar research.

Keywords: river engineering, vulnerability, RUSLE, soil erosion potential model (SEPM), soil erosion

1. Introduction

Soil erosion is a critical global issue that leads to significant land degradation and reduced soil productivity [1]. It involves the removal of soil particles primarily through the actions of water and wind. While some researchers focus exclusively on erosion caused by precipitation, others consider both natural and human-induced factors [2]. This phenomenon is a serious environmental concern, stripping productive topsoil and affecting stream channels associated with agricultural land, forestry, transport, and recreation [1–3]. Erosion has a lasting impact on soil productivity and sustainability.

Soil erosion is the detachment, transport, and deposition of soil particles by forces like water, wind, and human activity [4–6]. It involves three phases: detachment, transportation, and deposition. Key agents include biotic factors (water, ice, wind, organisms, and human actions) and chemical processes like corrosion [3, 6]. Precipitation and wind erosion are particularly impactful, affecting large areas and land use economics [5, 6].

Water erosion occurs when the shear stress of water exceeds soil particle strength, influenced by factors such as raindrop impact, velocity, depth, and slope gradient [3, 6, 7]. Soil and raindrop properties also play a role, but data on these is limited (**Figure 1**) [8–10].

Three types of soil erosion are identified: (1) micro-erosion, which includes sheet and small-scale rill erosion; (2) meso-erosion, involving rill and various lake or river erosions; and (3) macro-erosion, which refers to significant erosion from rivers and seas, such as gully and in-stream erosion [6, 8]. Understanding these processes and their influencing factors is crucial for effective erosion control in watershed management [2, 3, 11].

Dislodged materials are transported by raindrops or flowing water [12]. **Figure 2** illustrates how raindrop energy (e) and stream power (Ω) facilitate soil detachment and transport. Detachment occurs when energy (e) meets or exceeds the critical threshold (e_c), and flow transport occurs when stream power (Ω) exceeds the critical flow threshold Ω_c (loose) raindrop detachment-flow transport happens.

Deposition occurs when flow velocity decreases due to low slope, rainfall, surface roughness, or plant stems. The settling of particles is affected by their size, shape, and density [13, 14]. The processes of detachment, transport, and deposition are interconnected, as a particle may undergo these stages multiple times [15, 16]. Maintaining soil structure and implementing erosion control measures are essential for effective sediment management (**Figure 3**) [17, 18].

Observing soil erosion in isolation is challenging. A practical approach combines field observations with laboratory tests to simulate erosion over specific time scales and locations. However, estimating soil loss under varying conditions can be complex and time-consuming [3]. To address these issues, models can predict erosion in different scenarios. Various models exist, differing in processes, complexity, and data requirements. Model selection depends on its function, emphasizing reliability, global applicability, and suitability for diverse climates and land uses [3, 19]. Recommending

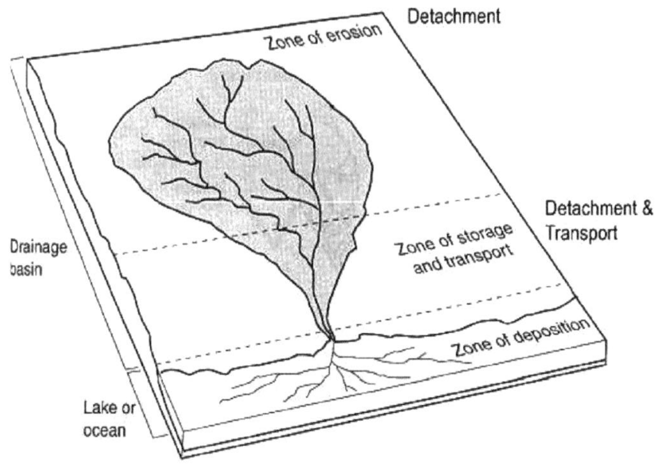

Figure 1.
Soil erosion procedure by water [8].

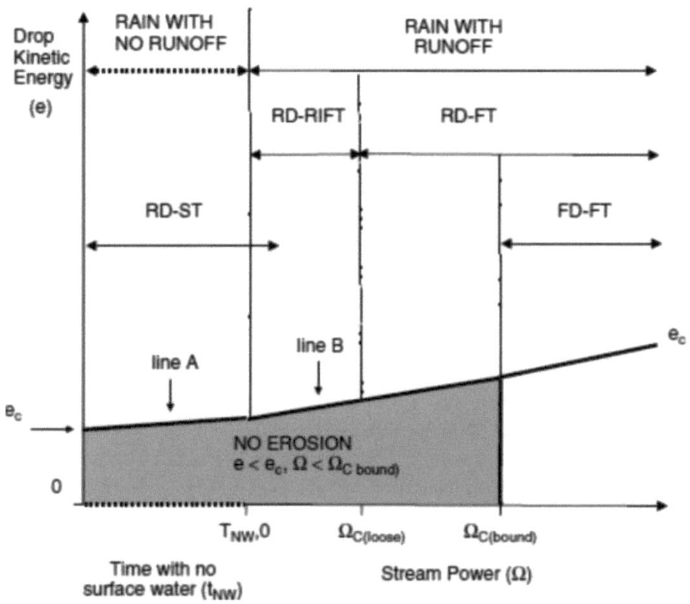

Figure 2.
Detachment and transport processes associated with variations in raindrops, by Kinnell [12].

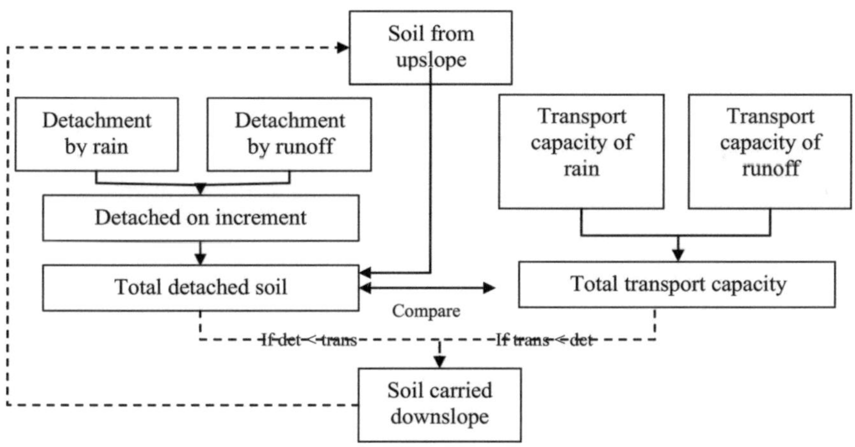

Figure 3.
Flow chart of the processes of soil erosion model with water [1, 3].

a single best model is impractical due to these differences. Key factors for model effectiveness include [20]:

- Objective and performance of the model
- Characteristics of the component include catchment, data, and expected output.

Meyer and Wishmeier [21] noted Zingg's early work on soil loss evaluation in 1940. Since then, several surveys have aimed to predict soil erosion. The three main types of models—empirical, conceptual, and physically based—often overlap; for instance, rainfall-runoff models can be either physical-based or conceptual, while empirical relationships can also predict erosion [21, 22].

Empirical models are the simplest soil erosion models, requiring limited data and calibrated parameters. They typically use stochastic analysis to predict average soil loss and identify sediment sources [5]. However, they rely on impractical assumptions about physical characteristics, precipitation, and soil type, often overlooking non-linearities, which has led to criticism [2, 21]. Despite these limitations, empirical models can still generate soil erosion criteria with partial data.

Conceptual models describe catchment characteristics without detailing specific procedures. They serve as a bridge between empirical and physically-based models [2, 21]. These models can assess the quantity and quality effects of land use changes without requiring detailed spatial and temporal input data [21, 23, 24].

Physical-based models depict soil erosion mechanisms by establishing relationships between factors and their spatial and temporal impacts [25]. They consist of two stages: first, generating a mathematical model that describes erosion processes, including detachment and transport by rainfall and flow. Second, these equations are solved using computer code, relying on point data to simulate erosion distribution. Small-scale parameters often represent larger-scale data, leading to a lack of theoretical justification, with no specific models for small catchments [5, 21]. **Table 1** summarizes the most suitable model types along with their applications and limitations.

Models are key tools for erosion analysis, involving (a) comparison with data, (b) sensitivity checks, and (c) accuracy tests [5, 26]. An ideal model is reliable, simple, comprehensive, and adaptable to changes in inputs like climate and land use [3, 20]. Thus, empirical and simple conceptual models are preferred [27].

- Require few data sets.
- Enhance understanding of erosion processes.
- Allow extrapolation beyond the data range and are user-friendly for stakeholders responsible for watershed management.
- Can combine empirical and conceptual models for more appropriate result.

This study will focus on the Revised Universal Soil Loss Equation (RUSLE) due to its relevance to erosion concepts and various simulation models for the following reasons:

- The required data is not too complex or inaccessible.
- It is straightforward to assign and understand functionally.
- It integrates well with GIS and remote sensing systems.

As a raster-based analysis, RUSLE predicts soil erosion potential on a cell-by-cell basis [1, 28–30].

The Universal Soil Loss Equation (USLE) was developed in the mid-1950s by Wischmeier, Smith, and their team at the National Runoff and Soil Loss Data Center, becoming widely used in the United States and globally [1]. In 1978, they introduced a revised version that included a soil erodibility nomograph and cover management factor derived from rainfall simulations [31, 32].

Model classification	Model name	Scale	Application	Limitation
Empirical	USLE	Plot	Croplands, rangeland and forest	It does not account for ephemeral gullies, problem with multiple land use, do not predict deposition and sediment yield, no spatial and temporal variation
	RUSLE	Plot	Croplands, rangeland and forest	It does not predict deposition and sediment yield
	MUSLE	Plot/large watershed	Croplands, rangeland and forest	Does not account for ephemeral gullies, problem with multiple land use, do not predict deposition and sediment yield, no spatial and temporal variation
	IHACRES_WQ	Small watershed	Croplands, rangeland and forest	Parameters values must be calibrated against observed data, there is no land surface erosion component to the model which predict sediment yield
Conceptual	HSPF	Watershed	Runoff, sedimentation load	It relies on calibration against field data, relatively large number of parameters
	AGNPS	Watershed	Runoff, gully erosion,	Large data requirement, models need calibration
	LASCAM	Watershed	Salt and water balance model	Parameters values must be calibrated against observed data
	SWRRB	Large complex rural watershed	Rainfall, runoff, subsurface flow	Large amount of component
Physically-base	ANSEWRS	Watershed	Croplands, ranglands	It is not capable for long-term simulation
	CREAM	Field sized area up to 40 ha	Gully erosion and deposition	The accuracy of the model highly related to accuracy of input, soil topography and land use assumed uniform
	TOPOG	Watershed up to 10 km^2	Rainfall-runoff model	It is complicated in structure, detail requirement which limit widespread application, do not predict sediment transport and channel erosion
	WEPP	Small watershed	Croplands, rangeland and forest	It does not account for gullies or permanent channels and streams, not suitable for large watersheds

Table 1.
Erosion sediment transport models.

In 1985, significant research led to the Revised Universal Soil Loss Equation (RUSLE), which retains the original USLE framework but features modified factor calculations applicable in various scenarios [30, 32]. Soil loss values are calculated based on RUSLE [32, 33]:

$$A = R.K.L.S.C.P \tag{1}$$

where:

A is the average of spatial or temporal soil loss per unit of area which is expressed in ton. ha-1. yr-1 (another unit such as mt. ha-1. yr-1 can be applied);

R is the rainfall-runoff erosivity factor in MJ. mm. ha-1.h-1. yr-1, which represents the energy and intensity of rainstorm.

K is a soil erodibility factor (t. ha. h. ha-1. MJ-1. mm-1), which is a measure of the erodibility of specific soil as measured under a standard condition of plot.

The form L and S are the dimensions for slope length and steepness, and C and P are dimensionless of cover management and erosion control practices. An important advantage of RUSLE is that it was represented by a committee of soil conservationists' scientists who had significant practice and knowledge of the soil erosion process [31]. Practically, the USLE/RUSLE has been the most commonly used model in estimating soil loss rates [14, 33].

According to Renard and Ferreira [34] and Galetovic et al. [32], a computer-based model was developed to estimate soil loss using RUSLE and organized into three database groups. This model predicts soil erosion on hillslopes from raindrop impact, overland flow, and rill erosion, applicable to various land uses such as mining and construction, and it can estimate deposition on concave slopes or dense vegetation.

Millward and Mersey [35] integrated RUSLE with GIS in a mountainous tropical watershed, demonstrating its ease of use and compatibility with GIS and remote sensing. Their study used data from digitized topographic, soil maps, and Landsat Thematic Mapper imagery (TM).

Tiwari et al. [36] compared the watershed erosion prediction project (WEPP) model with USLE and RUSLE across 20 natural runoff plots. WEPP, a computer-based model, offers benefits for spatial and temporal analysis of topography, erosion, and sediment detachment, but USLE and RUSLE showed greater accuracy in predicting soil erosion.

Pham et al. [37] applied RUSLE at a global scale using GIS to evaluate soil loss. They provided digital elevation models, soil properties, land cover, and precipitation data at a resolution of approximately 0.5°. Their findings offered a global perspective on soil erosion by water, noting that model accuracy is more related to dataset resolution than site size.

Walling [38] highlighted that RUSLE can predict erosion deposition and sediment load on concave slopes and terrace channels, suggesting changes in soil erodibility through regression equations.

The USDA (2003) [39] proposed a new model with distinct features compared to RUSLE, predicting sediment yield based on soil texture and being independent of land use. This model does not directly use RUSLE factors for erosion calculation but computes values within a mathematical framework [40].

Since 1994, various enhancements have been made to RUSLE. Water resource managers and environmental planners now prefer models that provide spatial distributions of soil erosion, integrating remote sensing and GIS to enhance soil conservation efforts [39–44].

The aim of this research is to develop a spatial soil erosion potential model (SEPM) to assess soil erosion risk, with a focus on both long-term and seasonal average potential. The study will also compare the SEPM results with those from similar catchments.

2. Materials and methods

2.1 Study area

The Ulu Kinta Catchment is situated between latitudes 4.29′–4.31’ N and longitudes 101.03′–101.12′ E. This narrow river valley features two right-angle bends and follows a northwest trend, covering approximately 30,752 ha. It is one of seven subdistricts in the Kinta Valley, which includes Sungai Kampar, Sungai Raya, Sungai Terap, Tg Tualang, Sungai Teja, and Sungai Belanja. The Kinta Valley is a broad alluvial plain that runs north to south, with a population density of around 940,000. The capital of Perak state, Ipoh, lies within this area, which is 1 of 14 major river basins in Malaysia. The basin is divided into six districts, totaling 15,180 km^2 (**Table 2**). The rivers originate in the northern mountains near Sungai Pari and flow southward to the Strait of Malacca. The catchment area is 14,700 km^2, covering nearly 70% of Perak State [45]. **Figure 4** illustrates the study area's location.

The geomorphology of the study area features lithological and alluvial units, with thicker alluvial coverage in the eastern part compared to the valley. The riverbed consists of coarse sand transitioning to large boulders [46]. The Sg Kinta flows along the western side of the Kinta Valley and has been canalized from Ipoh to Tanjung Tualang. Elevation ranges from 38 m in the west to 2160 m in the northern and eastern hilly areas. The profile of the west-east elevation trend in the study area is shown in **Figures 5** and **6**.

The climate in the study area is characterized by seasonal rainfall, consistent yearly temperatures, and high humidity. The most notable seasonal feature is the periodic change in prevailing winds. Rainfall patterns include two dry seasons (January, February, June, July, and August) and two wet seasons (March, April, May, September, October, November, and December). The average annual rainfall from 1975 to 2010 was 2500 mm, with a mean daily temperature of 27°C. Monthly temperatures remain stable, with a minimum of 22°C in January and a maximum of 35°C in March, averaging between 25° and 27°C. **Figure 7** illustrates the average precipitation and temperature distribution in the study area based on meteorological data.

2.2 Data sets

To apply the RUSLE for simulating soil erosion distribution in a catchment, it is essential to develop mapping and interpolation techniques to create a comprehensive

No	Name of distinct	Area (km^2)
1	Hulu Perak	6563
2	Kuala Kangsar	2541
3	Kinta	1985
4	Perak Tengah	1279
5	Batang Pandang	1651
6	Hilir Perak	1161

Table 2.
The Perak Basin district [45].

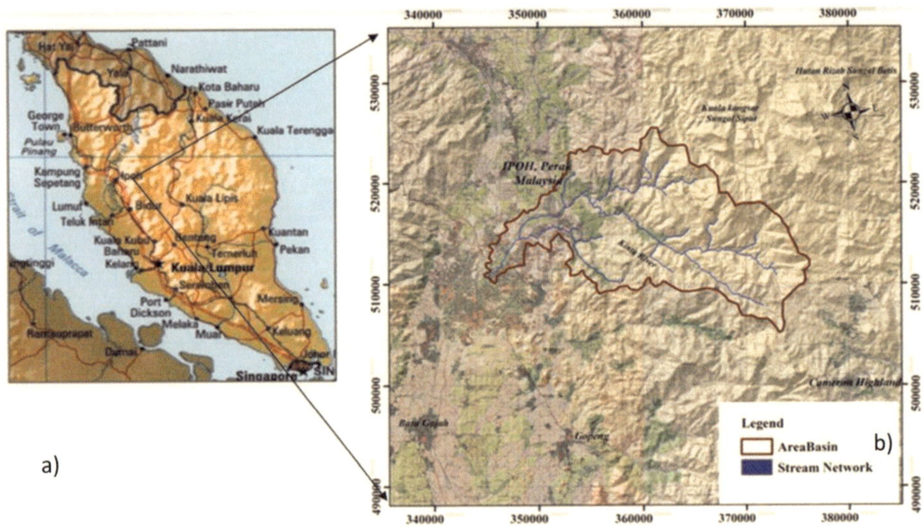

Figure 4.
(a) Location of the Ulu. Kinta catchment (Ipoh, Malaysia) and (b) topography map of the study area.

database [47]. This database must include all necessary inputs for the equation, using homogenous cells as small as possible for accurate erosion estimation. The available data for soil loss analysis include meteorological and Earth surface data.

Meteorological data, which have the longest history among satellite data, provide extensive temporal and spatial coverage. Key sources include the Global Precipitation Climatology Project (GPCP), the European Union's Global Climate Models (GCMs), MODIS, and the Tropical Rainfall Measuring Mission (TRMM) [48–52]. This study uses TRMM for the first time to estimate the spatial distribution of the rainfall erosivity factor.

Earth resources data reflect characteristics of features such as vegetation, soil, and water, defined by land use/land cover (LULC). Land use pertains to how humans utilize land and its resources, while land cover refers to the physical layers of human construction, natural cover, and vegetation [48]. Three satellites are introduced for this study: Shuttle Radar Topography Mission (SRTM), Landsat, and SPOT, which are considered the best for soil erosion estimation [49].

2.3 Method of soil erosion potential model (SEPM)

A conceptual framework was developed to estimate the spatial distribution of the soil erosion potential model (SEPM) based on the RUSLE model, integrating remote sensing (RS), GIS, and ancillary data. This model aims to determine the spatial distribution of soil erosion within the catchment and was chosen for the following reasons:

- Simple and attainable according to its effectiveness factor,
- Easy to implement and understand from a functional perspective,
- Compatible with RS and GIS data [2, 35].

To manage this purpose, ERDAS 9.1 and ArcGIS 10, helped to estimate soil loss rates. The results of the study will be represented (**Figure 8**).

Figure 5.
Overview of different LULC in Ulu Kinta Catchment, (a) urban area, (b) agriculture, (c) Barren land and water body, (d) forest, and (e) grassland and forest.

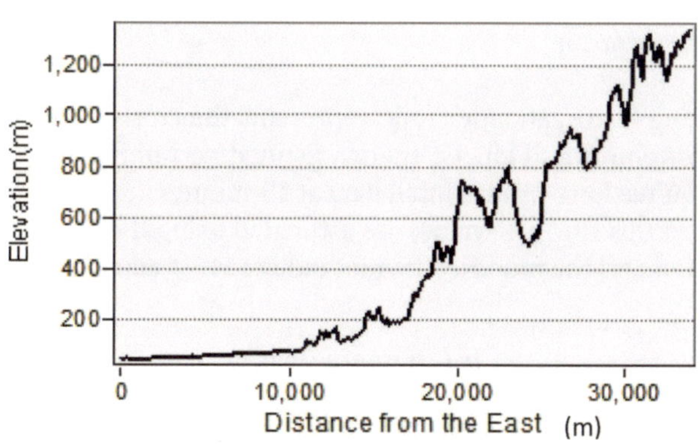

Figure 6.
Profile of elevation trend (west–east) in the study area.

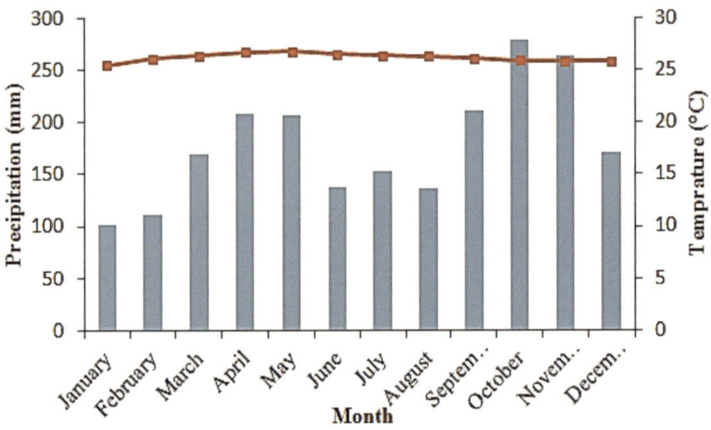

Figure 7.
Average monthly precipitation and temperature distribution for the Ulu Kinta Catchment.

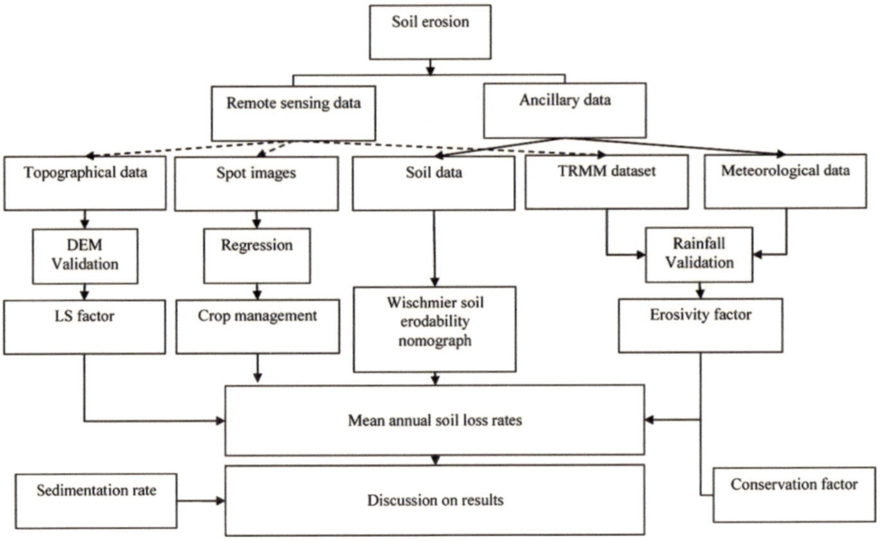

Figure 8.
Soil erosion potential model (SEPM).

2.3.1 Rainfall erosivity factor

The precipitating factor, an index unit, represents the erosive force of specific rainfall, defined by its volume and kinetic energy. Estimating rainfall erosivity based on kinetic energy requires long-term rainfall data at 15-minute intervals. Due to inadequate pluviograph data in this study, R-values are estimated using the Morgan equation (2) for annual precipitation and the modified furnier index (MFI) equation (3) for seasonal analysis, that is,

$$R = 9.28P - 8835.15 \tag{2}$$

where R is rainfall erosivity MJmm ha-1 h-1 yr-1 and P is annual mean precipitation in mm. For the present analysis, the R-factor was estimated from possible available rain gauge data, as the catchment has no record of daily rainfall intensity:

$$R = 38.5 + 0.35P \tag{3}$$

where R is rainfall erosivity MJmm ha-1 h-1 yr-1 and P is annual mean precipitation in mm. This equation in Asia is well known as the EI-Swaify et al. equation [50].

2.3.2 Meteorological precipitation data analysis

2.3.2.1 Rain gauges data

To calculate the value for P, a precipitation distribution prediction procedure is employed. The primary model uses a correlation matrix to explain precipitation patterns through multivariate nonlinear regression, assessing how topographical factors influence precipitation. ERDAS 9.1 software was utilized to determine topographical variables, extracted from a 90 m resolution SRTM DEM with a project coordinate system of RSO-Malaysia (**Figure 9**) in a "natural" setting.

2.3.2.2 Tropical rainfall measuring Mission dataset (TRMM)

Data were collected using monthly TRMM 3B43 (0.25° × 0.25°) from 2000 to 2010 in 4-byte real binary, accessible from NASA (**Figure 10**). For more accurate precipitation from TRMM downscaled results, incorporating auxiliary data like topography and vegetation index is beneficial for catchment-scale applications (**Figure 11**), [52, 54].

A multivariate sharpening technique was applied to NDVI, DEM, and TRMM 3B43. NDVI and TRMM were first resampled to 1 km resolution. Then, a polynomial regression was used to establish the relationship between precipitation and NDVI, yielding the best fit for the observed data in Eq. (4):

$$P = b_0 + bX_1 + bY_2 + b_3NDVI + b_4NDVI^2 \tag{4}$$

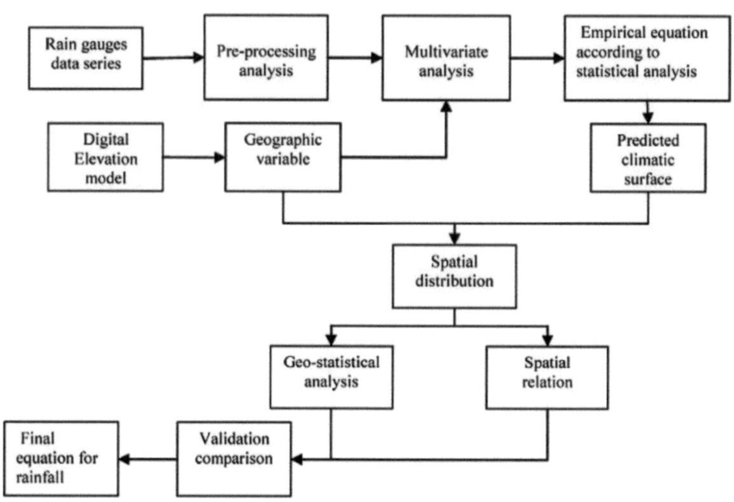

Figure 9.
Diagram of the rain gauges analysis developed for the study.

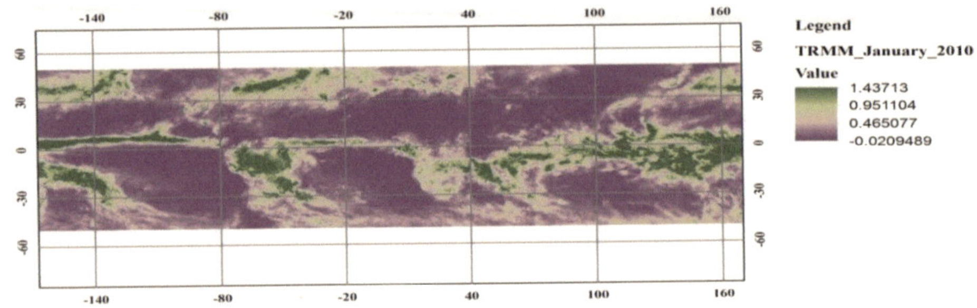

Figure 10.
TRMM 3B43 projected [53].

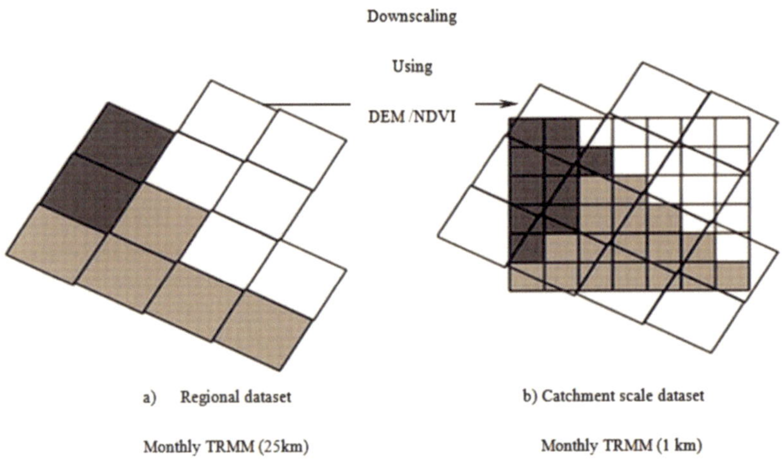

Figure 11.
Reduce resolution of TRMM3B43.

where P represents precipitation (mm/h), b0 remark a constant; b1 .... b4 are coefficients obtained for each independent variable, and NDVI represents the biomass index:

$$P = b_0 + b_1X + b_2Y + b_3DEM + b_4DEM^2 \quad (5)$$

All variables are structured similarly to Eq. (4). While, In this study, instead of utilizing the NDVI index, we employed Digital Elevation Models (DEMs) to analyze locations (X, Y).

To evaluate how the models predict precipitation, a three-stage verification procedure was conducted, that is (**Figure 12**).

i. A comparative analysis was performed by validating the monthly downscaled results from the DEM regression model against rain gauge data for mean annual, dry, and wet seasons from 2000 to 2010. The downscaled precipitation regression based on NDVI was verified using rain gauge data for corresponding dates.

ii. Correlation coefficients (R2) for both station and predicted data were assessed to further validate methods.

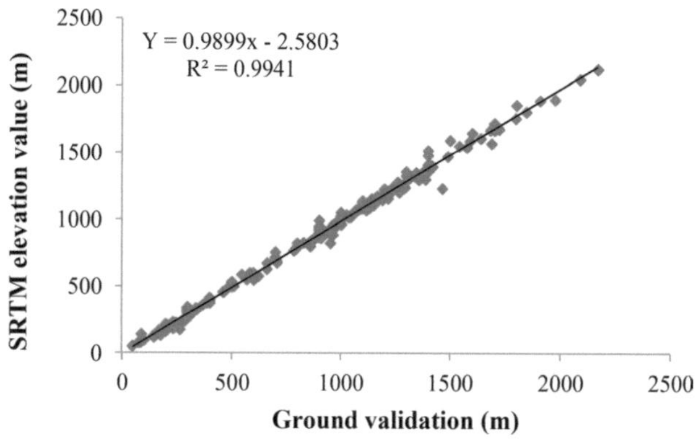

Figure 12.
Validation result of SRTM and topographic map.

iii. Bias was calculated using Eq. (6) to determine whether each prediction model overestimated or underestimated precipitation,

$$B = \frac{\sum_{i-1}^{n} M_1}{\sum_{i-1}^{n} O_i} - 1 \tag{6}$$

where M is the TRMM 3B43 or predicted precipitation, O is the observed precipitation from the gauges, I, n are the station number index and total number of gauges.

Finally, model performance was statistically compared using root mean square error (RMSE) and mean absolute error (MAE) (**Figure 13**).

2.3.3 Soil erodibility factor (K)

Soil erodibility refers to the soil's resistance to detachment and transport. In the catchment's thick forested regions, soil survey efforts faced limitations. A coarse sampling approach was adopted using a generalized soil texture map derived from the 2010 DOA soil map of Malaysia, assigning corresponding soil types. The sampling procedure and K factor determination were adapted from [55]. Tew [56] based on the Eq. (7)

$$K = \left[1 \times 10^{-4}(12 - OM)M^{1.14} + 4.5(s - 3) - 8(p - 2)\right]/100 \tag{7}$$

where M is the % silt + % very fine sand, multiplied by (100-%clay), OM is the % organic matter, s is the soil structure code based on **Figures 14** and **15**, P is the permeability code as a function of S (soil structure) (**Table 3**).

2.3.4 Slope length and slope steepness (LS)

Soil erosion is influenced by slope length and steepness. To assess the impact of topography on erosion, the LS factor and SRTM DEM were analyzed. A validation

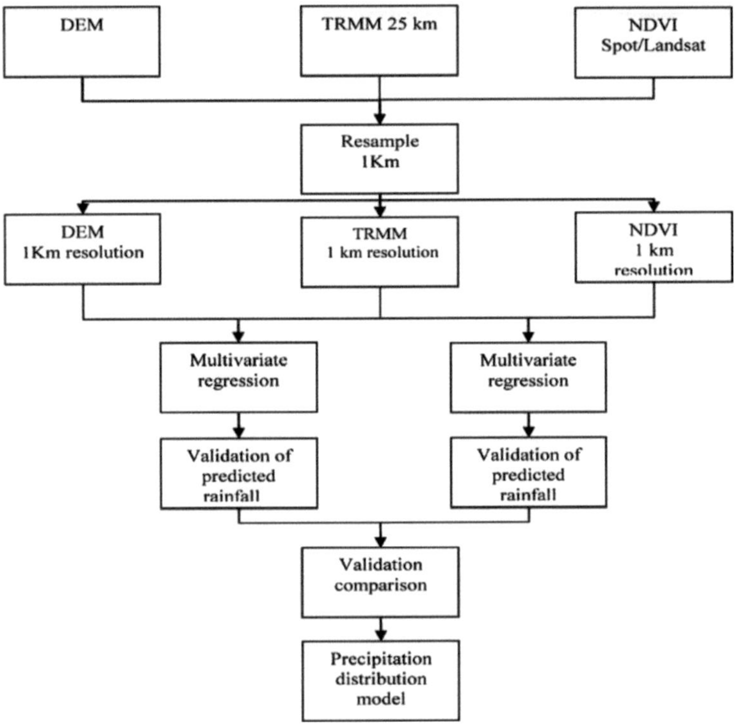

Figure 13.
TRMM analysis procedure.

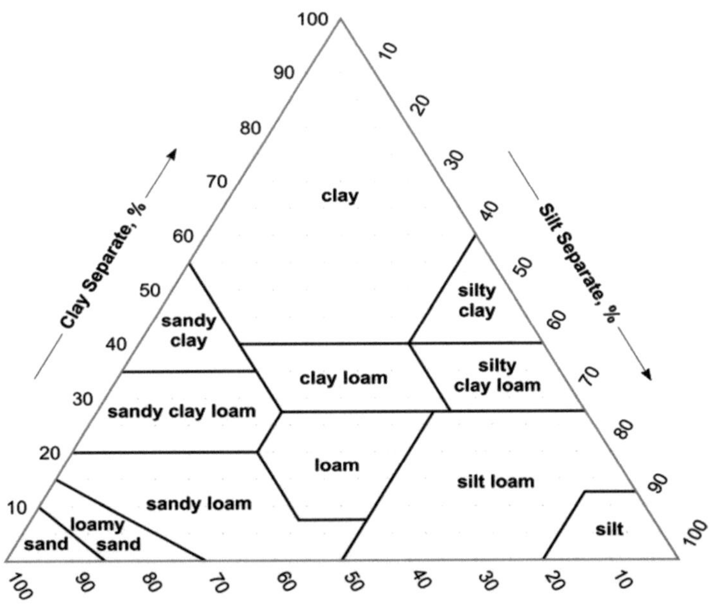

Figure 14.
Determining s value using USDA [47].

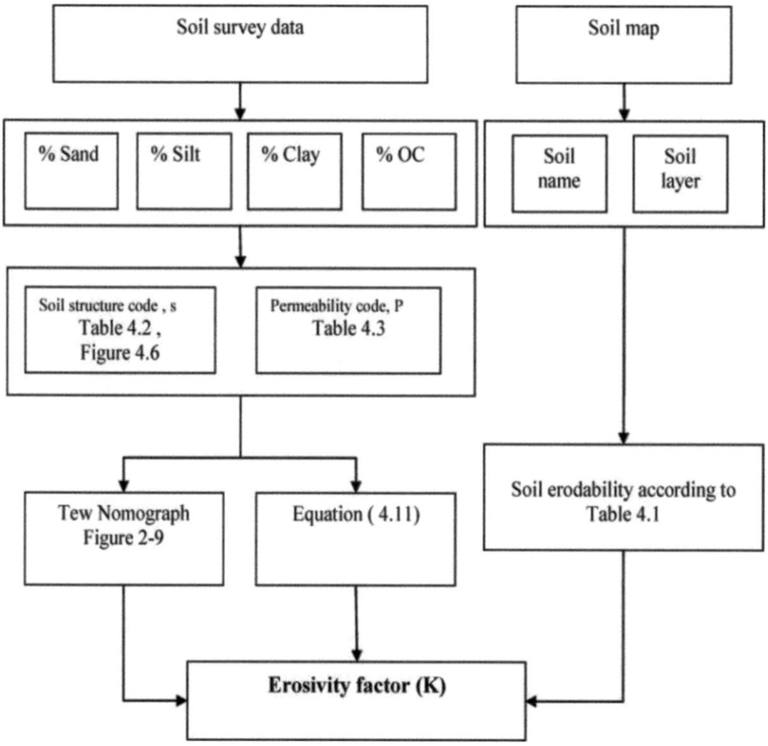

Figure 15.
K factor determination procedure.

Texture		Permeability code	Hydrologic soil group
Sand	Fast	1	A+
Loamy sand, sandy loam	Fast to moderately fast	2	A
Loam, silt loam	Moderately fast	3	B
Sandy clay loam, clay loam	Moderately fast to slow	4	C
Silty clay loam, sandy clay	Slow	5	C-D
Heavy clay, clay	Very slow	6	D

Table 3.
Soil permeability data for major soil textural classes [31].

procedure used a 1:50,000 scale topographical map and 200 ground elevation points for accuracy enhancement. A linear regression showed a 99.4% correlation with the ground validation points (**Figure 12**). The SRTM 90 m data was considered suitable for calculating LS. A filling sink was conducted to reduce discontinuities, followed by determining flow direction and accumulation. **Figure 16** presents the LS factor layer requirements.

2.3.5 Management factor (C and P factor)

Crop management and practice factors (C and P factors) are crucial for controlling soil loss rates in a given region. Currently, there are no specific values for these factors

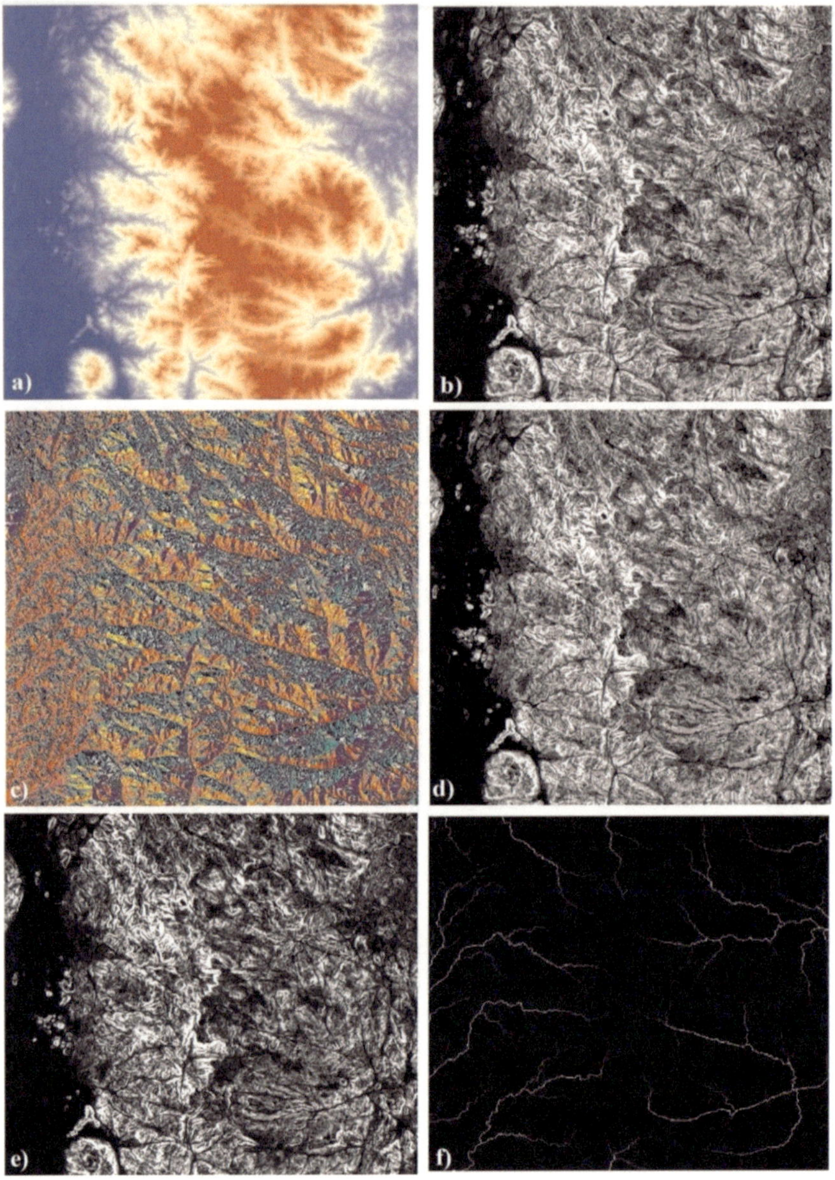

Figure 16.
Schematic view of LS factor calculation steps; (a) DEM, (b) slope, (c) flow direction, (d) m factor, (e) S index, and (f) L factor (flow accumulation).

in the study area. To determine the C and P factors, satellite imagery was utilized to define the vegetation index and land use/land cover (LULC). Image analysis is essential for generating accurate and interpretable remote sensing images, serving as a key source for visual interpretation and data extraction. The initial steps for image processing are outlined in the following sections (see **Figure 17**).

The C factor in RUSLE represents the impact of vegetation—such as trees, grass, and cropland—on soil erosion, influencing soil loss rates significantly. To estimate the C factor cell by cell at a resolution of 3 m × 3 m, a relationship between NDVI values

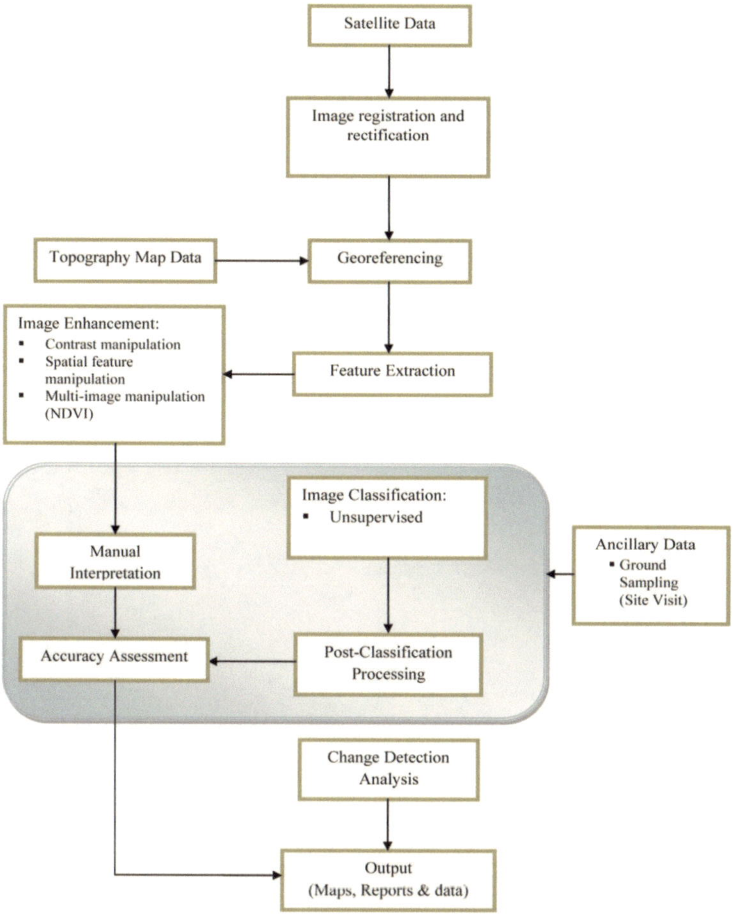

Figure 17.
Image processing procedure.

and corresponding C values from the RUSLE guidelines was established. The C value is 0 for well-protected or dense forest areas and set at 1 for bare soil. To assign C values, 30 samples from forest areas with NDVI greater than 0.4 and 30 samples from bare soil classes with NDVI less than 0.2 were extracted from each NDVI image. A linear regression analysis was then conducted using IBM SPSS 19 to determine the relationship between NDVI and C values.

Finally, the C factor was distributed across the study area by extrapolating the equation using the raster calculation tool (Kriging) in ArcGIS 10. This process included removing shadows and clouds from the NDVI image, coding them as no data, and applying an average process with smoothing to a 3 × 3 pixel size (Eq. (8)):

$$C = 1.131 - 1.472NDVI \quad (8)$$

where C is covering management and NDVI is a normalized difference vegetation index.

The optimal resolution of the RUSLE model in this study is 90 m; consequently, the distribution map of the C value was adjusted to a resolution of 90 m × 90 m (**Figure 18**, **Table 4**).

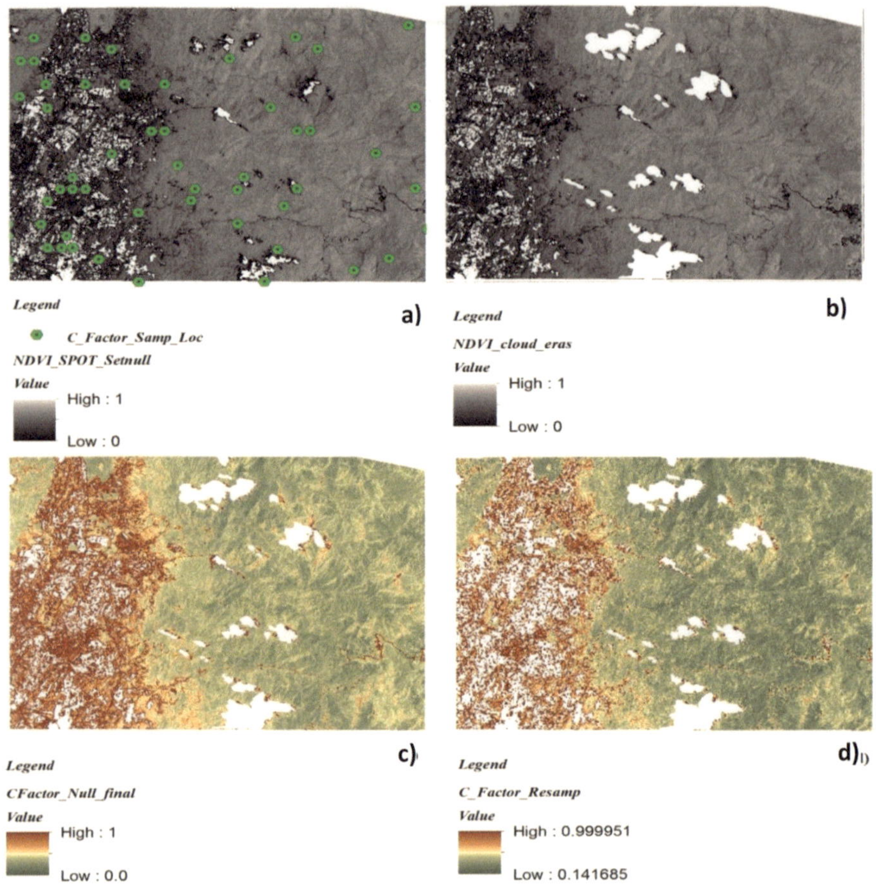

Figure 18.
Estimating C factor according to NDVI: (a) NDVI and sampling location, (b) removing shadow and clouds, (c) estimated C factor, and (d) resampled C factor from 3 m to 90 m resolution.

The erosion control practice factor (P) represents the effectiveness of surface actions on runoff. In this study, conservation practices were defined through P value classification based on land use-land cover (LULC) results extracted from SPOT5_2010 images using ERDAS 9.1 software. To improve the clarity of the LULC image, clouds and shadows were removed, and the P values were distributed using Kriging methods to fill in any gaps. Since the P factor is influenced by contouring and land shaping, the final P values were assigned by reclassifying the slope map according to **Table 5**, and the slope map was reclassified. Ultimately, the determination of the P factor was based on a multi-criteria fuzzy overlay of the slope layer and LULC data layer, following the approach outlined by Sunil Saha et al. [54]. The fuzzification procedure establishes the distribution possibilities of the practice value, as illustrated in **Figure 19**, which defines a finite set of points referred to as elements, along with their properties as

$$A = \{x, \mu_A(x)\} \text{ for each } x \in X \tag{9}$$

where membership x (the object) in A and $x \in X$indicates that x is an object of contained in X and practically it defines as $X = \{x_1, x_2, \ldots, x_n\}$, and $\mu_A(x)$ is between 0 and 1.

	Descriptive factors		Statistic	Std. Error
C _Factor From sampling point	Mean		.5586	.04859
	95% Confidence Interval for Mean	Lower Bound	.4606	
		Upper Bound	.6567	
	5% Trimmed Mean		.5622	
	Median		.3900	
	Variance		.102	
	Std. Deviation		.31860	
	Minimum		.03	
	Maximum		1.00	
	Range		.97	
	Interquartile Range		.64	
	Skewness		.557	.361
	Kurtosis		−1.298	.709
Estimated C factor	Mean		.5417	.04475
	95% Confidence Interval for Mean	Lower Bound	.4513	
		Upper Bound	.6320	
	5% Trimmed Mean		.5376	
	Median		.3800	
	Variance		.086	
	Std. Deviation		.29342	
	Minimum		.09	
	Maximum		.99	
	Range		.90	
	Interquartile Range		.60	
	Skewness		.621	.361
	Kurtosis		−1.319	.709

Table 4.
Descriptive analysis of regression model.

2.3.6 Spatial distribution of soil erosion potential model (SEPM)

The potential soil loss rates, representing the geo-environmental scenario of the study area, were analyzed using the Spatial Analyst tool in ArcGIS [55]. The five-factor layers were transformed into grid formats with 90 × 90 m cells, all aligned in the same coordinate system. These GIS input layers were then multiplied using the RUSLE to estimate mean annual and seasonal soil loss rates on a pixel-by-pixel basis, creating a spatial distribution model for soil erosion potential.

Support / Sediment control practice		P factor
Bare soil		1.00
Disked bare soil (rough or irregular surface)		0.90
Wired log/ Sand bag barriers		0.85
Check dam		0.80
Grass buffer strips (to filter sediment laden sheet flow)		
Basin slope %		
0 to 10		0.60
11 to 24		0.80
Contour furrowed surface (maximum length refers to down slope length)		
Slope %	Max length	
1 to 2	120	0.60
3 to 5	90	0.50
6 to 8	60	0.50
9 to 12	40	0.60
13 to 16	25	0.70
17 to 20	20	0.80
>20	15	0.80
Silt fence		0.55
Sediment containment systems (sediment basin trap)		0.50
Berm drain and cascade		0.50
terracing		
Slope %		
1 to 2		0.12
3 to 8		0.10
9 to 12		0.12
13 to 16		0.14
17 to 20		0.16
>20		0.18

Table 5.
P Support practice, P factor.

2.3.7 Validation of SEPM model

To verify the output of the SEPM based on the RUSLE model, we conducted a validation procedure with two distinct scenarios. First, we analyzed the results of each factor individually to assess the accuracy of the remote sensing data. This step helps identify any specific factors that might influence the model's output and ensures the reliability of the data used. Second, we compared the SEPM outputs with previous studies conducted in areas with similar geo-environmental and rainfall characteristics.

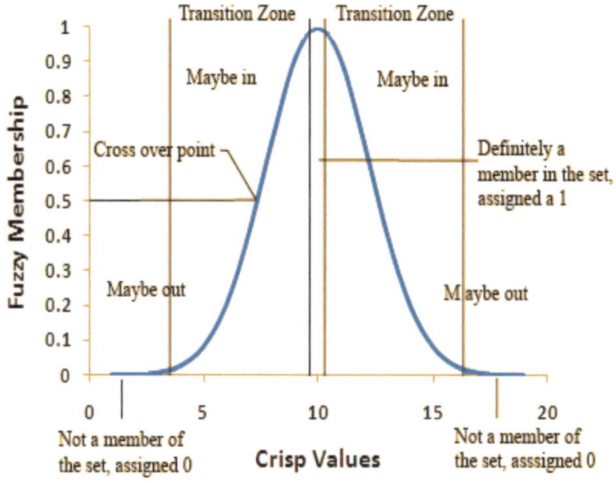

Figure 19.
Fuzzy membership function diagram [54].

This comparison provides a broader context for validation and helps confirm the model's applicability across comparable environments.

2.4 Result and discussion

2.4.1 Results

2.4.1.1 Rainfall erosivity (R)

Among nine rain gauges, five initially had missing data. To address this, monthly rainfall data were aggregated to fill gaps and ensure consistency. A double mass curve analysis was conducted from 1974 to 1994, and a modification factor was calculated to estimate missing data points (**Table 6**) [57, 58].

The relationships between precipitation and independent factors were analyzed in a "natural" setting, leading to the development of four scenarios based on associated variables, as shown in **Table 7**.

Model 4 demonstrated the highest response rates for precipitation prediction, achieving 74%, 78%, and 63% for mean annual, dry season, and wet season precipitation, respectively. The spatial distribution of average annual precipitation was found to correlate with elevation, slope, and aspect. This interpolation was conducted using a combination of multivariate regression and geospatial interpolation, employing an exponential model. A lag size of 12 km was established, resulting in a total of 12 lags. In this model, b_0 represents the constant, while b_1 to b_4 are the coefficients for each independent variable (X and Y). The DEM (Digital Elevation Model) indicates the location (X and Y) and altitude (m):

$$\begin{aligned} P = b_0 + bX_1 + bX_2^2 + bX_3^3 + bY_4 + bY_5^2 + bY_6^3 + \\ b_7DEM^1 + b_8DEM^2 + b_9DEM_6^3 + b_{10}Slope^1 + \\ b_{11}Slope^2 + b_{12}Slope^3 + b_{13}Aspect^1 + b_{14}Aspect^2 + b_{15}Aspect^3 \end{aligned} \quad (10)$$

Month	Index gauge	Correction factor							
		R1	R2	R3	R4	R6	R7	R8	R9
January	R5	1.23	1.16	0.79	1.9	1.23	0.68	0.9	0.97
February	R5	1.05	1.07	1.114	1.33	1.18	1.69	0.88	0.92
March	R5	1.0	0.82	1.10	[h]1.01/ 0.93	0.94	1.47	1.04	1.20
April	R5	0.84	1.045	094	0.98	0.86	1.06	0.92	1.02
May	R5	0.91	1.37	1.27	1.16	1.26	1.0	1.15	0.98
June	R5	0.84	1.08	1.5	0.95	0.98	0.93	0.86	1.03
July	R5	1.05	1.07	1.103	0.93	[b]0.74/ 1.09	0.86	0.89	0.8
August	R5	0.85	1.27	1.21	1.12	1.18	1.03	0.98	1.12
September	R5	0.78	0.91	[a]1.05/ 0.96	0.89	1.02	[d]0.76/ 0.88	[f]0.71/0.89	1.3
October	R5	0.83	0.96	1.13	0.96	[c]1.08/06	1.25	1.02	0.95
November	R5	0.84	1.02	1	0.89	0.88	[e]1.05/1.09	[g]0.74/ 0.89	1.67
December	R5	1.03	1.03	1.04	0.64	0.94	1.22	1.14	1.17

[a]*Before 1982/after 1983.* [b]*Before 1983/after 1984.* [c]*Before 1988/after 1989.* [d]*Before 1979/after 1980.* [e]*Before 1988/after 1989.* [f]*Before 1980/after 1981.* [g]*Before 1982/after 1983.* [h]*Before 1985/after 1988.*

Table 6.
Modification factor in rain gauges based on the double mass curve.

Rend surface and 3D plot analyses provided valuable insights into the distribution patterns of annual and seasonal precipitation (see **Figure 20**). The trend analysis from 1974 to 2010 indicated a steady increase in annual average precipitation from the west to the northeast. During the dry season, precipitation was mainly concentrated in the western region, while lower values were observed in the highlands. In contrast, the wet season saw higher precipitation levels in the central part of the study area. To enhance the capabilities of Model 4, the spatial distribution of precipitation patterns was considered for the annual average, dry season, and wet season.

This interpolation combines multiple regression and geospatial interpolation, utilizing an exponential model. After a trial procedure, a lag size of 12 km was fixed, resulting in a total of 12 lags. **Figure 21** illustrates an example of advanced surface modeling. The digital elevation map (**Figure 21(a)**) enhances the understanding of elevation patterns and their relationship with precipitation. Trend analysis reveals a similar behavior, with even clearer distributions evident across three scenarios. The spatial distribution of annual average precipitation shows a correlation with elevation, slope, and aspect, indicating a gradual increase in rainfall from west to northeast. This suggests the presence of three distinct zones within the distribution map.

To assess the accuracy of the estimated statistical and spatial models, we considered the root mean square error (RMSE), mean absolute error (MAE), and mean absolute deviation (MAD). The results of this analysis are presented in **Table 8**. The most accurate result was observed during the wet season, with a correlation

DOI: http://dx.doi.org/10.5772/intechopen.1008237

Model	Topographical variable	Average				Dry season				Wet season			
		ANOVA Result	R^2	Adj-R^2	SE	ANOVA Result	R^2	Adj-R^2	SE	ANOVA Result	R^2	Adj-R^2	SE
1	Rain, DEM, Slope	0.051	0.201	−0.07	9.59	0.08	0.052	−0.264	5.39	0.06	0.136	−0.15	18.16
2	Rain, DEM, Slope, Aspect	0.05	0.352	−0.31	9.47	0.058	0.305	−0.112	5.06	0.059	0.296	−0.13	17.96
3	Rain, DEM, Slope, Aspect, location	0.08	0.420	−0.55	11.57	0.051	0.328	0.031	4.72	0.09	0.313	−0.83	22.90
4	Eq. 2 variables	0.03	0.740	0.721	21.79	0.073	0.63	0.51	3.86	0.041	0.781	0.702	16.68

Table 7.
Models summary characteristic.

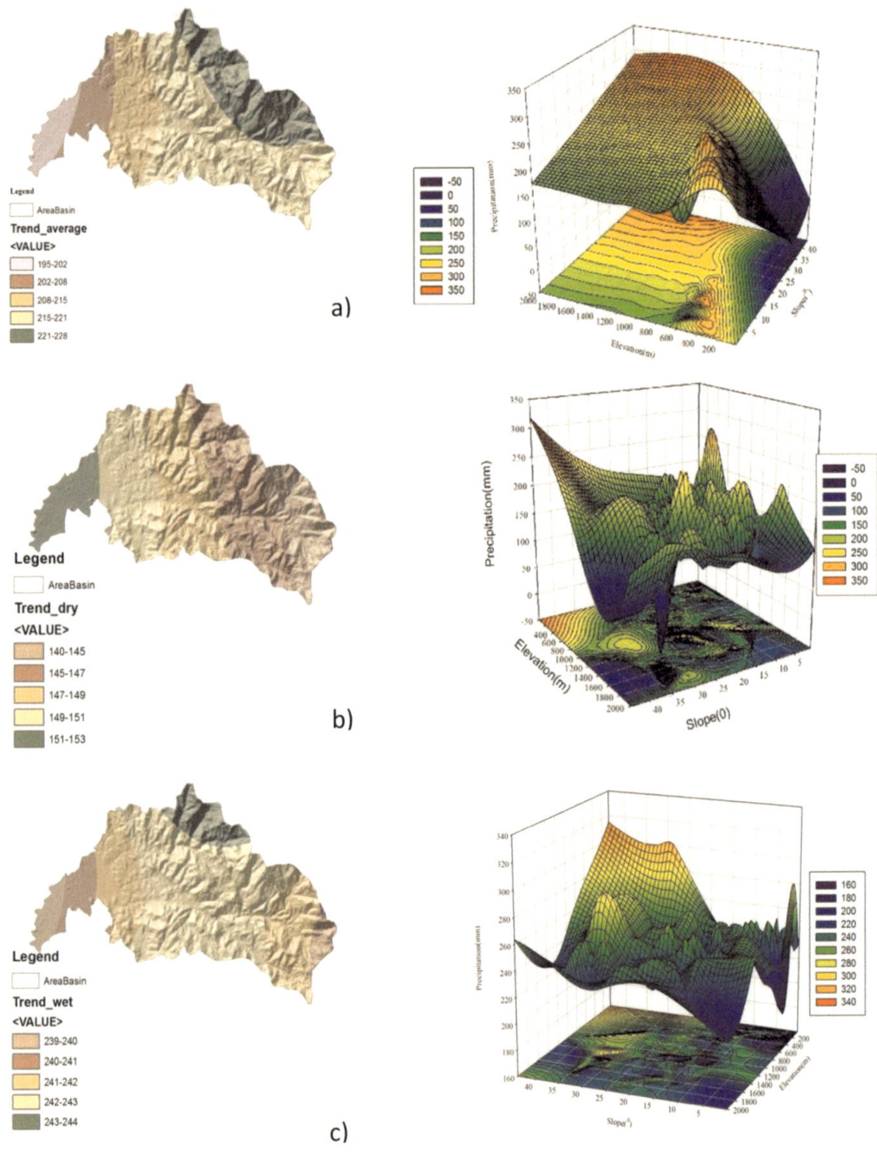

Figure 20.
Precipitation trend (mm/month) in the study area by surface map and 3D plot, (a) annual average rainfall trend, (b) dry season, and (c) eet season trend.

coefficient of 0.78. In contrast, the statistical analysis yielded response rates of 0.74 for the annual average and 0.63 for the dry season. Additionally, the coefficient of determination improved across all cases after applying the spatial analysis model.

To obtain high-resolution information, a new downscaling algorithm was performed by using DEM and NDVI images [59]. **Figure 22** shows the NDVI SPOT (1 km) image taken in February 2000 alongside TRMM 3B43 precipitation at 0.25° resolution. The scatter plot of downscaled TRMM 3B43 and NDVI (Landsat TM+ and SPOT) with the regression line is depicted in **Figure 23**. The initial analysis included 10 TRMM precipitation samples across three categories (mean annual, dry, and wet seasons), with dry and

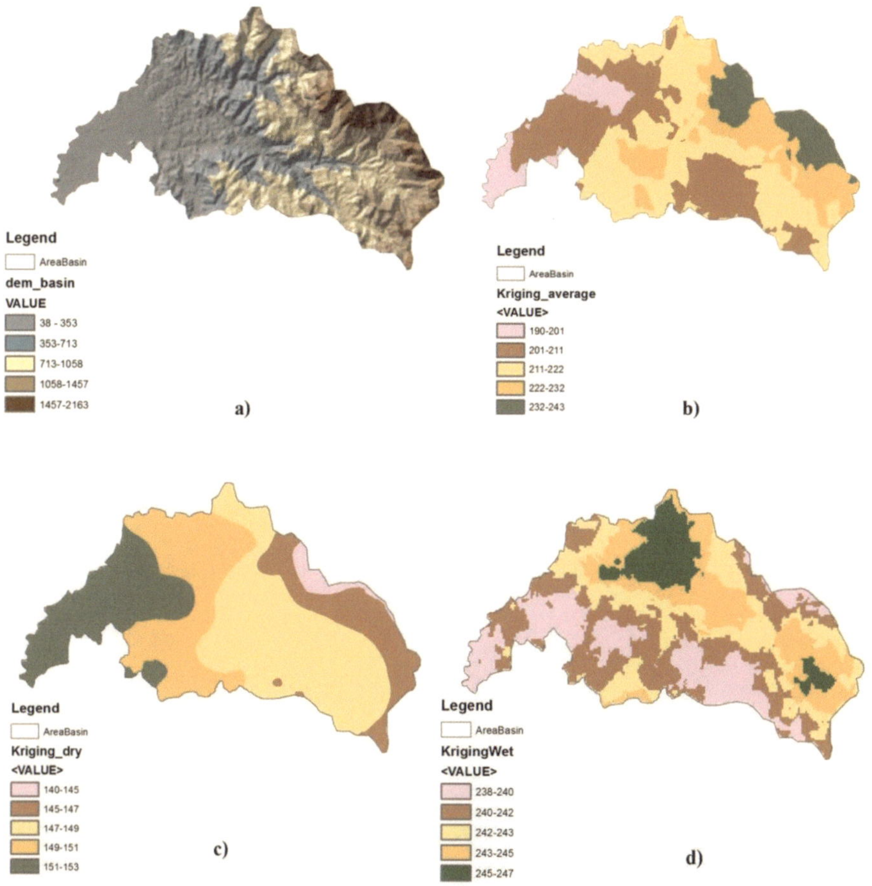

Figure 21.
O'Kriging interpolation methods. (a) digital elevation model of the study area, (b) annual average precipitation distribution (mm/month), (c) precipitation distribution in the dry season (mm/month), and (d) precipitation distribution in the wet season (mm/month).

Model	Regression results				Geospatial analysis (kriging Model)			
	R2	RMSE (mm)	MAE (mm)	MAD (mm)	R2	RMSE (mm)	MAE (mm)	MAD (mm)
Annual	0.74	21.88	24.32	11.23	0.88	19.38	17.79	10.97
Dry season	0.63	20.58	24.85	15.87	0.80	23.77	14.97	9.19
Wet season	0.78	26.38	21.58	13.46	0.89	19.52	15.81	13.2

Table 8.
Accuracy measurement for precipitation models, Mean, RMSE, MAE and MAD.

wet months established earlier. The graphs indicate a relationship between precipitation and NDVI, with fitting function R^2. However, the preliminary analysis revealed a weak correlation: R^2 values ranged from 0.08 to 0.72 for NDVI Landsat and from 0.008 to 0.50

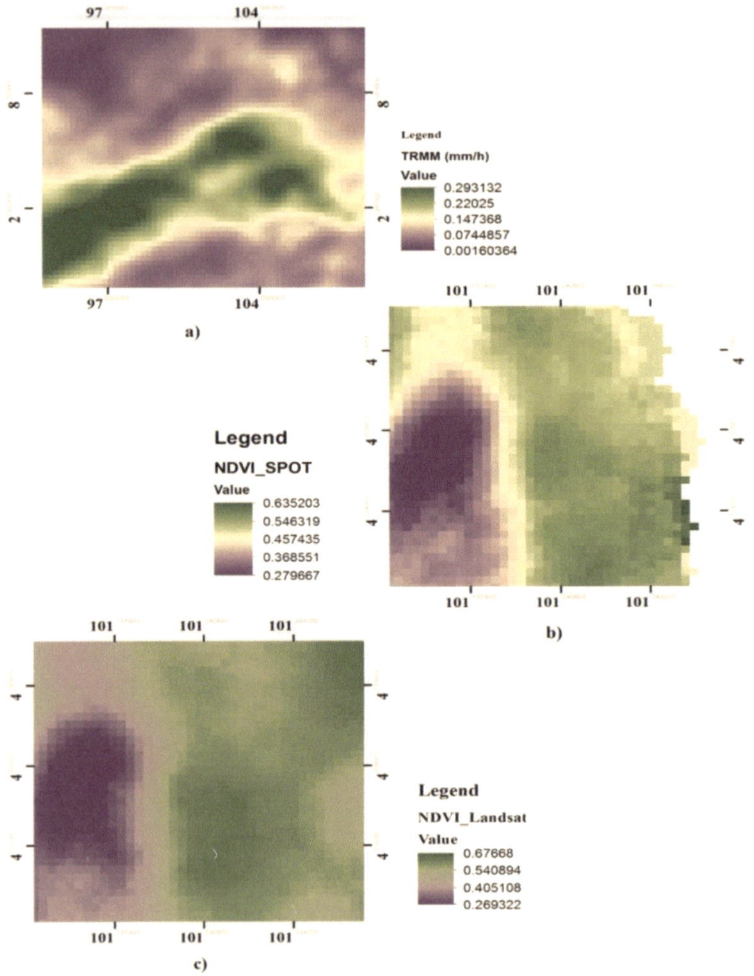

Figure 22.
(a) Average TRMM 3B43 precipitation (0.25° resolution) in February 2000 (mmh^{-1}), (b) $NDVI_{SPOT}$ (1 km resolution) in February 2000, and (c) $NDVI_{Landsat}$ (1 km resolution) in January 2005.

for NDVI SPOT. Huffman et al. [60] and Sorooshian et al. [54, 57, 61] noted that the NDVI and TRMM relationship is imperfect due to orbital deviation and retrieval algorithm flaws. Consequently, an alternative method was employed to enhance this relationship based on earlier assumptions in Eqs. (4 and 5).

The mean annual precipitation from TRMM 3B43 and DEM for 2000 to 2010 at 0.25° resolution is shown in **Figure 24**. Monthly precipitation ranged from 0.06 to 0.5 mm/h. A monthly statistical downscaling procedure was performed, as described in Section 3-2-2, to extract seasonal rainfall stratification for estimating rainfall erosivity.

To illustrate the relationship between TRMM precipitation and elevation, a scatter diagram for three categories (mean annual, dry season, and wet season) is presented in **Figure 25**. The figure indicates a significant relationship between precipitation and DEM, with preliminary analysis showing correlation coefficients between 0.5 and 0.78. However, a limited dataset revealed a weaker relationship during the dry season in 2010, suggesting that the connection between elevation and precipitation is not very strong. To achieve more accurate results, the analysis was refined based on Eq. (5).

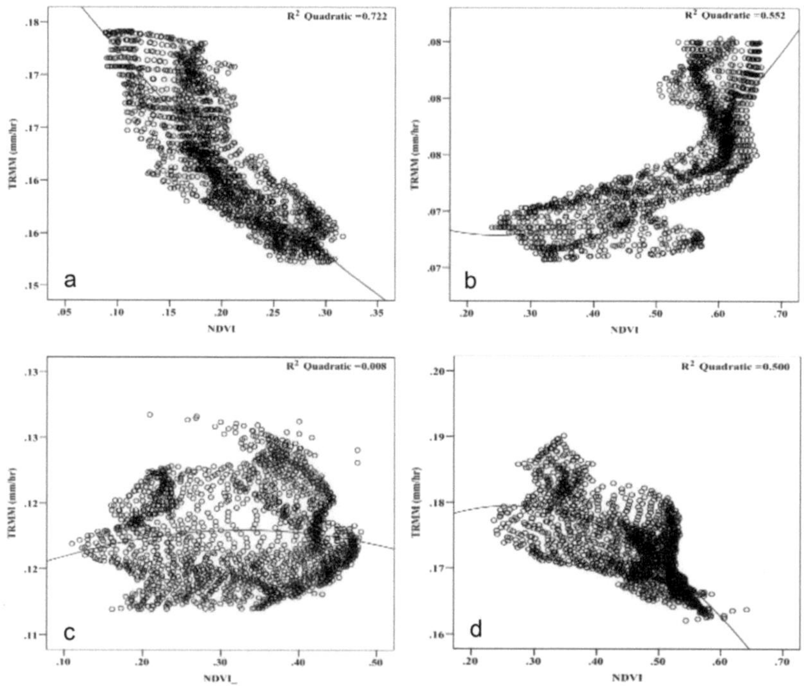

Figure 23.
Scatter diagram between TRMM 3B43 (mmh^{-1}) and NDVI for dry month and wet month in sample images (Landsat TM and SPOT), (a) Landsat TM (21/9/2001)_TRMM, (b) Landsat TM (26/1/2005)_TRMM, (c) SPOT (10/9/2004)_TRMM, and (d) SPOT (20/2/2010)_TRMM.

A multivariate regression model utilizing TRMM 3B43 and DEM data from 2000 to 2010 is shown in **Figure 26**. The model's coefficients for mean annual, dry, and wet seasons were significant at $P \leq 0.05$, with high determination coefficients (R^2 ranging from 0.93 to 0.99 and adjusted R^2 from 0.873 to 0.983). Despite some overestimations and underestimations, the predicted rainfall trends aligned closely with observed data. The accuracy assessment of the spatial distribution model demonstrates that pluviograph data and the downscaled TRMM 3B43 dataset, adjusted by DEM, can effectively estimate the erosivity factor for soil loss rates. Consequently, rain gauge data and the downscaled TRMM 3B43 with DEM were used to generate rainfall erosivity, as shown in **Table 9**. According to the accuracy assessment, to evaluate rainfall erosivity, two data series were considered. First, was rain gauge data, and second was downscaled TRMM3B43 based on DEM.

In this part, the result of annual and seasonal rainfall erosivity analysis at nine pluviographic stations is presented. The seasonal scale was considered to reproduce the seasonal effect of the precipitation pattern, which displays the spatial distribution of rainfall erosivity based on the pluviograph dataset: (a) mean annual, (b) wet season, and (c) dry season, showing R-values from the distribution model. The erosivity results demonstrate spatial and temporal variations influenced by the elevation model. Mean annual erosivity ranged from 12,433.54 to 18,269.85 MJ mm ha^{-1}h^{-1}yr.$^{-1}$, with a mean value of 15,189.67 MJ mm ha^{-1} h^{-1} yr.$^{-1}$ (**Figure 27**).

Medium erosivity levels covered 27.17% and 21.95% of the study area during the mean annual and wet seasons and 30.37% in the dry season. Additionally, 33.4% and 47.3% of the region exhibited the lowest rainfall erosivity, characterized as below average and low, primarily located in the west and southwest at lower elevations.

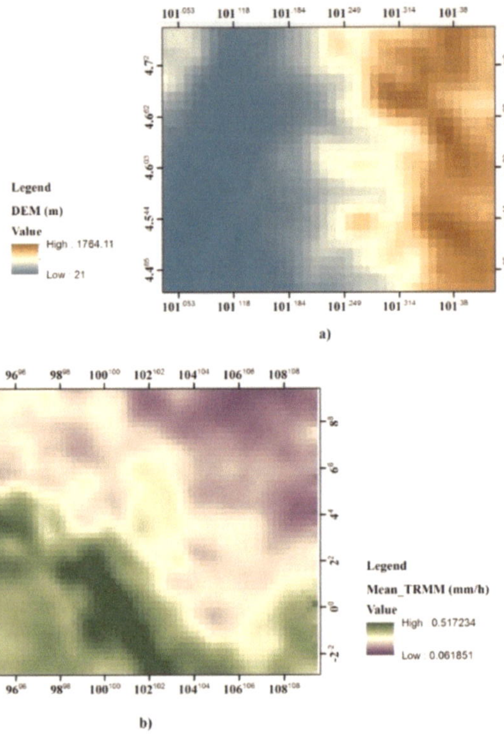

Figure 24.
(a) DEM (1 km resolution) and (b) TRMM 3B43, 0.25° resolution for long average rainfall (LAR) from 2000 to 2010.

Figure 28 Rainfall erosivity was illustrated using the downscaled TRMM 3B43 dataset. The minimum, maximum, and mean values of rainfall erosivity aligned closely with the results from the pluviograph data. Additionally, the TRMM dataset revealed clearer trends in erosivity changes compared to the absolute values.

Overall, both scenarios (rain gauge data and downscaled TRMM dataset based on DEM) yielded satisfactory results. The temporal and spatial changes in rainfall erosivity during the mean annual and wet seasons showed no significant differences between the rain gauges and TRMM datasets, with a clear trend distributed at 1 km resolution. While the frequency distribution of rainfall erosivity based on the TRMM 3B43 dataset was appropriate, caution is advised in using the TRMM modeling for the wet season due to poor correlation with precipitation (Dense forest cover in the catchment limited the soil survey organization, leading to coarse sampling based on a generalized soil texture map. **Table 10** shows the k-factor values from coarse soil sampling integrated into GIS, with spatial distribution results derived using inverse distance weighting (IDW). **Figure 29** illustrates the k-factor magnitudes and spatial distribution of soil types. K values range from 0.029 to 0.066 (ton/ha/MJ mm ha^{-1} h^{-1}). The steep eastern land and urban/mining areas in the west exhibit the highest erodibility potential. Factors such as topographical position, slope steepness, and rainfall significantly influence soil loss rates. In steep areas, silt presence and high-intensity rainfall increase erodibility, while in urban and mining regions, human activities and soil properties are key contributors.) Notably, the trend of rainfall erosivity estimated for the dry season diverged from that based on rain gauge data. Thus, the distribution indicates that rainfall erosivity can be estimated using TRMM 3B43, but attention must be given to wet and dry seasons.

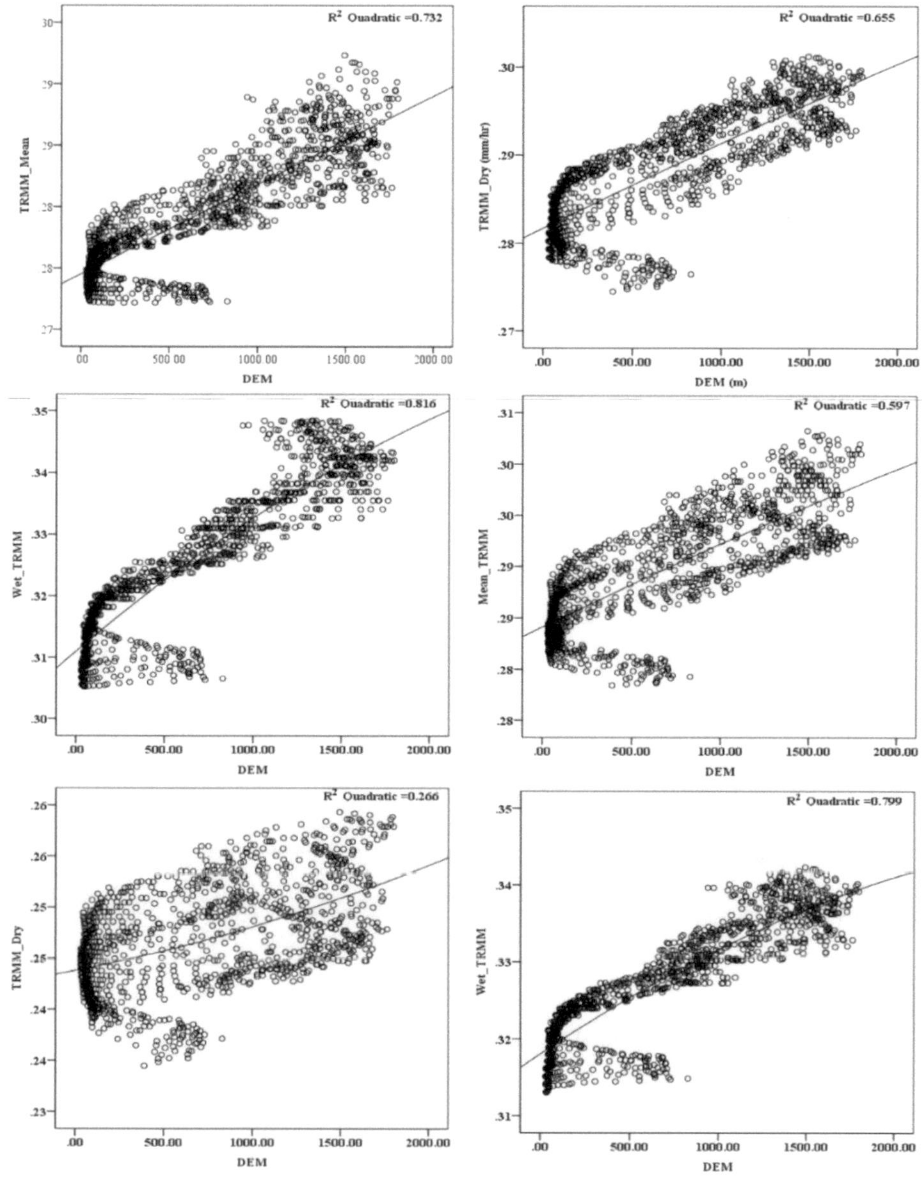

Figure 25.
Scatter diagram of TRMM 3B43 (mmh^{-1}) and DEM (m) for mean annual, dry season and wet season in sample year, (a–c) represent the year of 2000 results and (d–f) illustrate the year of 2010 results.

Utilizing monthly and annual data from rain gauges with a multivariate regression function for topographical variables allows users to assess the significance of interpolated values. Ultimately, incorporating rain gauge data with the topographical variable model enhances the resolution and accuracy of results from the TRMM 3B43 dataset, particularly for wet and dry seasons.

The models' robustness was confirmed by comparing them with published erosivity data (Table 5.14). Annual rainfall erosivity ranged from 5500 MJ·mm·ha^{-1}·h^{-1}·yr.$^{-1}$ in Cameron Highlands to 21,600 MJ·mm·ha^{-1}·h^{-1}·yr.$^{-1}$. Seasonal erosivity values varied between 554 and 659 MJ·mm·ha^{-1}·h^{-1}·yr.$^{-1}$. The proposed R-factor estimation

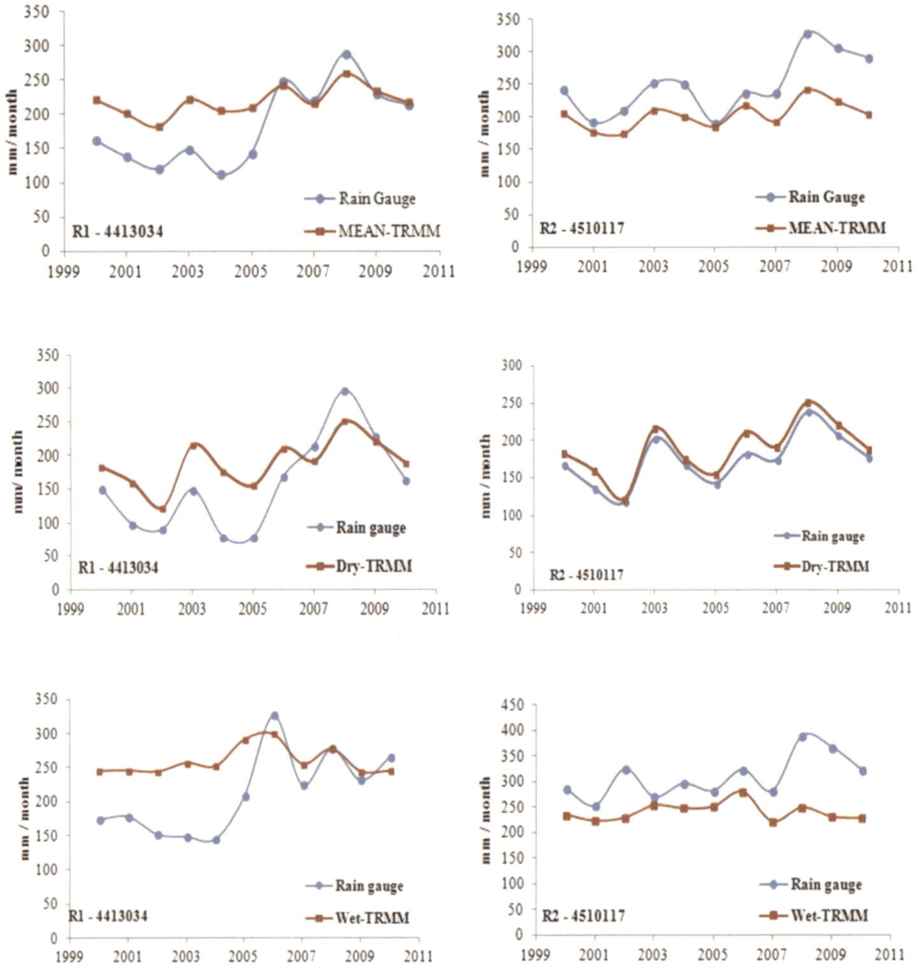

Figure 26.
Typical results comparing seasonal observed precipitation with TRMM 3B43 precipitation simulated via *the DEM downscaling algorithm.*

technique was deemed suitable for the study area based on rain gauge data. However, the seasonal TRMM dataset yielded inconsistent results due to its 1 km resolution.

2.4.1.2 Soil erodibility (K)

Dense forest cover in the catchment limited the soil survey organization, leading to coarse sampling based on a generalized soil texture map. **Table 11** shows the k-factor values from coarse soil sampling integrated into GIS, with spatial distribution results derived using inverse distance weighting (IDW). **Figure 29** illustrates the k-factor magnitudes and spatial distribution of soil types. K values range from 0.029 to 0.066 (ton/ha/MJ mm ha^{-1} h^{-1}). The steep eastern land and urban/mining areas in the west exhibit the highest erodibility potential. Factors such as topographical position, slope steepness, and rainfall significantly influence soil loss rates. In steep areas, silt presence and high-intensity rainfall increase erodibility, while in urban and mining regions, human activities and soil properties are key contributors.

		Scenario	R (%)	B	RMSE (mm)	MAE (mm)	MAD (mm)
	Rain Guage	Mean Annual	78.4	0.37	19.38	17.79	10.97
		Dry Season	65.25	0.15	23.77	14.97	9.19
		Wet Season	80.3	0.06	19.52	15.81	13.28
TRMM3B43	DEM	Mean Annual	77.4	0.03	10.8	3.24	2.22
		Dry Season	80.0	0.14	10.3	3.11	1.27
		Wet Season	28.3	−0.01	15.03	4.54	4.54
TRMM3B43 / NDVI	Landsat	Mean Annual	73.48	−0.27	22.48	6.77	10.97
	SPOT	Mean Annual	48	0.155	46.06	13.88	20.64

Table 9.
Accuracy results for precipitation models.

2.4.1.3 Slope length and slope steepness (LS)

Figure 30 illustrates the estimated LS factor in the Ulu Kinta Catchment. The LS factor distribution indicates that 75.3% of values range from 0 to 4, primarily in the western, southwestern, and middle areas. About 14.65% fall between 4 and 12, scattered across the mountainous regions in the north and south. Only 10.05% of the LS values range from 12 to 143, mostly in hilly areas. Overall, areas with greater slopes exhibit higher LS values, particularly along drainage borders.

2.4.1.4 Management factor (C and P)

The spatial distribution of the C factor in the study area, obtained through regression and geospatial analysis (Kriging) to fill gaps. The C values range from 0.21 to 0.954. Low C factor values are predominantly found in the eastern part of the catchment, where dense forest covers the area. Conversely, high C values are located in the western part, associated with bare soil or urban areas exhibiting low NDVI and significant human disturbance.

This analysis demonstrates that remotely sensed data can effectively estimate C factor values, providing essential information for enhancing C factor distribution at high spatial resolution. However, it is important to note that the C factor rarely reaches zero in practice, as soil loss can still occur during high-intensity rainfall, even when the surface is fully vegetated. Therefore, field surveys should be conducted alongside these remote-sensing images for more accurate assessments.

The effectiveness of terracing on soil erosion is integrated into the LS factor. The P factor was developed using land use/land cover (LULC) data extracted from SPOT 5 (2010), which identified seven major categories, with forests comprising 82% of the area. **Figure 31(a)** displays the assigned P values for each LULC type, indicating that 0 values represent areas with no data. **Figure 31(b)** shows the slope classification of the study area, as detailed in **Table 5**, while **Figure 31(c)** illustrates the distribution of P values across the catchment prior to slope assignment. Finally, **Figure 32(d)** presents

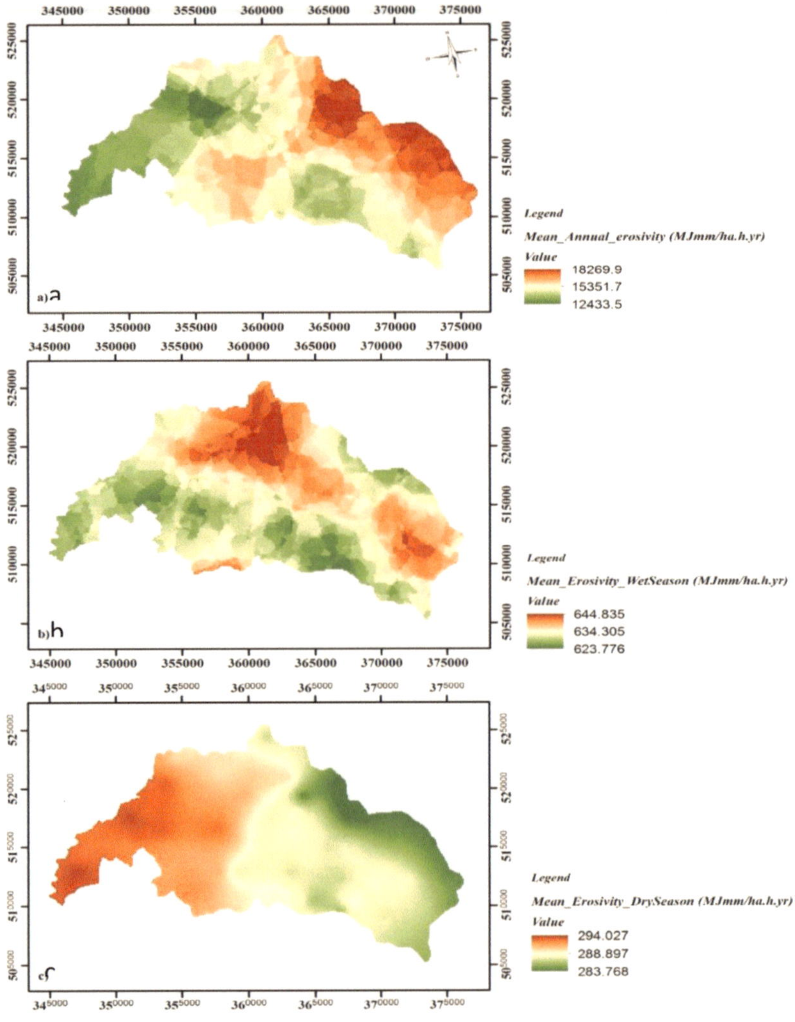

Figure 27.
Annually and seasonally of the spatial distribution of rainfall erosivity according to the pluviograph dataset, (a) mean annual, (b) wet season, and (c) dry season.

the estimated P values for each LULC, processed in ArcGIS 10. P factor values range from 0.5 to 0.8, with the highest values found in areas lacking conservation practices, particularly within forested regions. In contrast, the lowest P values are associated with urban areas, grasslands, and barren land. This trend indicates that higher P values reflect less effective conservation practices, underscoring the need for improved slope management and conservation strategies to mitigate soil erosion (**Figure 33**).

2.4.1.5 Spatial distribution of soil model

2.4.1.5.1 SEPM on the basis of rain gauges result

The potential soil loss rates, representing the geo-environmental scenario of the study area, were analyzed using the Spatial Analyst tool in ArcGIS [55]. The five-

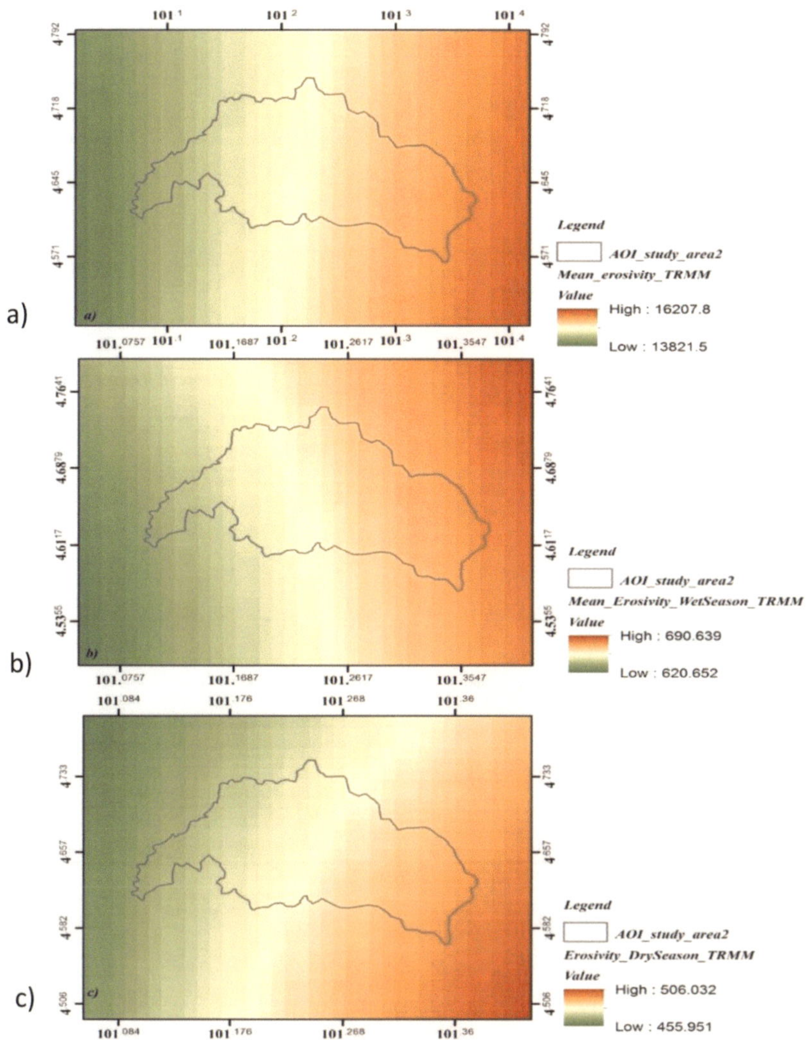

Figure 28.
Erosivity distribution value (%) in five categories (low, below average, average, above average, and high (MJmmha-1 h-1 yr-1), (a) mean annual, (b) wet season, and (c) dry season.

factor layers were transformed into grid formats with 90 × 90 m cells, all aligned in the same coordinate system. These GIS input layers were then multiplied using the RUSLE to estimate mean annual and seasonal soil loss rates on a pixel-by-pixel basis. This process created a spatial distribution model for soil erosion potential, which illustrates the average soil erosion rates estimated over 11 years, from 2000 to 2010. The predicted soil loss rates were categorized into six ordinal classes, detailing spatial distribution changes in erosion risk. According to these categories (**Table 12**), 80.9% of the area exhibited very low to low soil loss potential, while only 0.11% was classified as having extremely severe erosion risk. The mean annual soil erosion potential was calculated at 34.72 tons ha^{-1} $yr.^{-1}$, with a standard deviation of 40.93 tons ha^{-1} $yr.^{-1}$. The peak annual soil erosion value in a small region in the north reached 786 tons ha^{-1} $yr.^{-1}$, whereas most of the study area fell within the 150 tons ha^{-1} $yr.^{-1}$ erosion risk level.

Area (references)	Country	Method	Rainfall erosivity ($MJmmha^{-1} h^{-1} yr.^{-1}$)
[62]	Malaysia	E = 11.87 + 8.73log I	13,600–21,600
Penang [63]	Malaysia	$EI30 = 6.97rain_{10}-.23days_{10}$	9000–14,000
Cameron highland [64]	Malaysia	$R = 2.5P^2/100 (0.073P + 0.73)$	5500–9500
Ulu Kinta [65]	Malaysia	R = 9.28P-8838.15	10,326–14,233
Guidline for erosion and sediment control (Perak state) [56, 66]	Malaysia	$E = \sum_{r=1}^{k} e_r V_r$	10,000–20,000
[67]	Thailand	R = 38.5 + 0.35P	554.75–659.75
Ulu Kinta (Present Study)	Malaysia	Mean value: R = 9.28P-8838.15	
Rain gauge data			Mean annual: 12433–18,269
		Seasonally effect: R = 38.5 + 0.35P	Wet season: 623–644
			Dry season: 283–294
Ulu Kinta (Present Study)			Mean Annual: 13821–16,207
TRMM 3B43 dataset			
			Wet season: 620–690
			Dry season: 455–506

Table 10.
Comparison result of rainfall erosivity in different area.

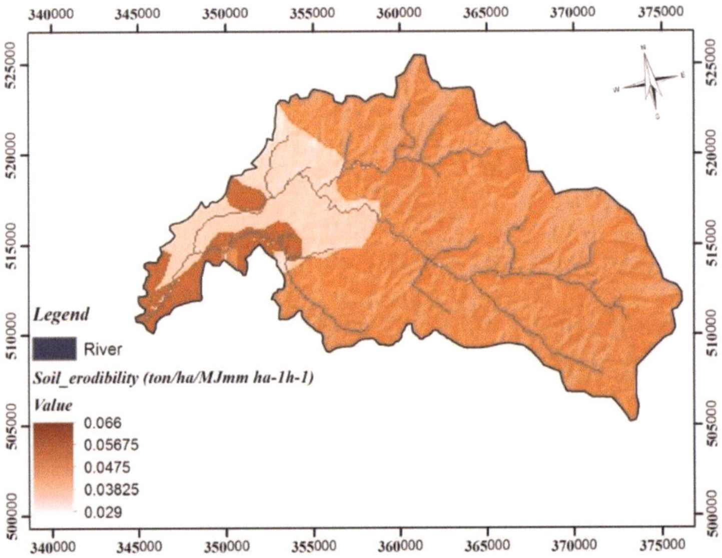

Figure 29.
Soil erodibility factor derived.

Sample	X	Y	Clay%	Silt%	Fine sand	OM%	S	P	K_s	K_{DOA}
1	353377.16	518557.16	9.64	58.23	16.8	6.14	4	3	0.035	0.027
2	358395.00	519774.06	15.6	42.14	12.8	4.3	4	3	0.034	0.056
3	360900.50	515544.91	22.1	12.2	56.4	7.12	4	4	0.041	0.056
4	357045.31	516436.38	0	65.4	3	12	3	4	0.016	0.029
5	355347.34	517092.09	17.4	45.12	11.8	14.23	3	5	0.027	0.029
6	352950.13	516713.44	4.61	71.3	6	3.7	3	3	0.039	0.035
7	348912.25	516649.59	4.61	68.3	10	4.6	3	3	0.036	0.035
8	347665.16	515262.09	1.5	76.5	6	4.7	4	3	0.043	0.035
9	349292.91	514406.16	1.3	73.7	8	5.1	4	3	0.041	0.066
10	346882.63	513486.5	0.93	51.97	4	16.2	4	4	0.019	0.035
11	345734.72	512265.91	0.94	58.36	6	12.1	3	4	0.021	0.066
12	346523.66	511361.97	0	58.4	3	11.7	4	3	0.017	0.066
13	348047.25	512260.25	29.5	52.1	4	14.4	4	5	0.033	0.066
14	351405.63	517491.59	23.2	2.54	47.3	6.18	4	4	0.039	0.066

Table 11.
Sampling result from fieldwork and calculated K value.

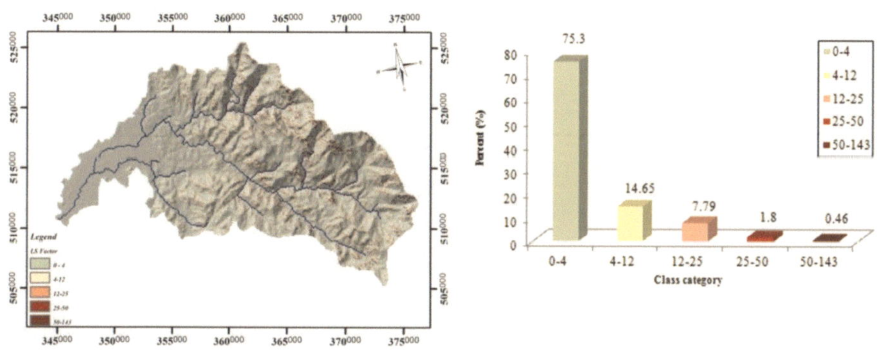

Figure 30.
LS factor distribution in the study area.

2.4.1.5.2 SEPM on the basis of TRMM erosivity factor

Precipitation erosivity was calculated using the TRMM dataset at a resolution of 1 km. Subsequently, soil loss potential was assessed with a raster calculator at a spatial resolution of 1 km × 1 km, both annually and seasonally.

Figure 34 Soil erosion potential (ton. ha^{-1} $yr.^{-1}$) on the basis of three scenarios of rainfall erosivity condition according to the TRMM dataset for the (a) mean value of soil erosion in dry season, (b) mean annual soil erosion, and (c) mean value of soil erosion in wet season shows that the average annual soil loss rate (SLR) ranges from 0.1 to 135.68 tons ha^{-1} $yr.^{-1}$, with an average of 7.9 tons ha^{-1} $yr.^{-1}$ and a standard deviation of 17.65 tons ha^{-1} $yr.^{-1}$. The spatial pattern indicates that SLR potential is highest in the east and northeast, where severe erosion occurs due to hilly terrain and high soil erodibility. In

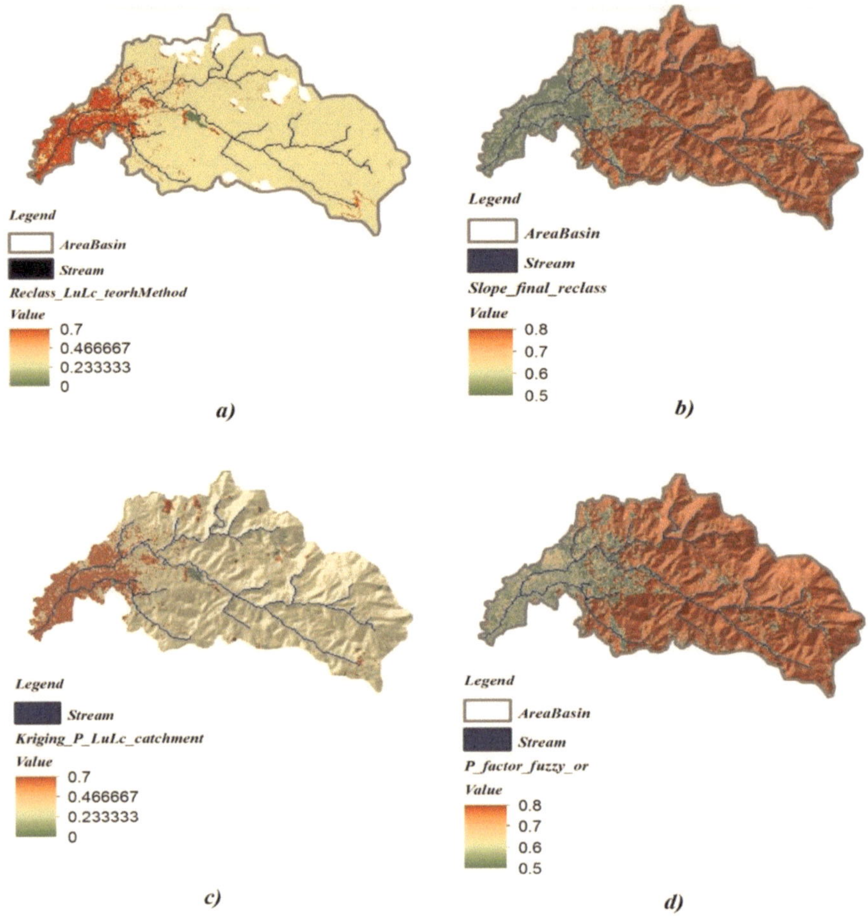

Figure 31.
Assigned P value factor to LULC image, (a) reclassified LULC, (b) reclassified slope, (c) P value without slope effect, and (d) countering effect on P value.

contrast, very low to low soil erosion rates are found in the eastern and central parts of the catchment, which are flatter and experience less rainfall erosivity.

The results emphasize that rainfall erosivity, soil erodibility, and the LS factor are the most significant contributors to SLR in this study. Overall, 97% of the study area falls into a category of low potential for annual soil erosion, and seasonal analyses show very low to low soil loss risk across the watershed. However, the 1 km × 1 km resolution of the TRMM dataset is considered too coarse for catchment-scale assessments. Despite this, the findings effectively highlight areas susceptible to erosion and provide realistic estimates of critical spatial patterns within the catchment (**Table 13**).

2.4.1.6 Validation of SEPM model

The evaluation of each factor was verified individually. Comparing our findings with previously published studies is essential for assessing the accuracy of the results, especially given the lack of comparable fieldwork characteristics in this area. **Table 14** presents soil loss rates (SLR) from published studies alongside relevant geo-environmental and rainfall characteristics. The validation procedure indicates that the

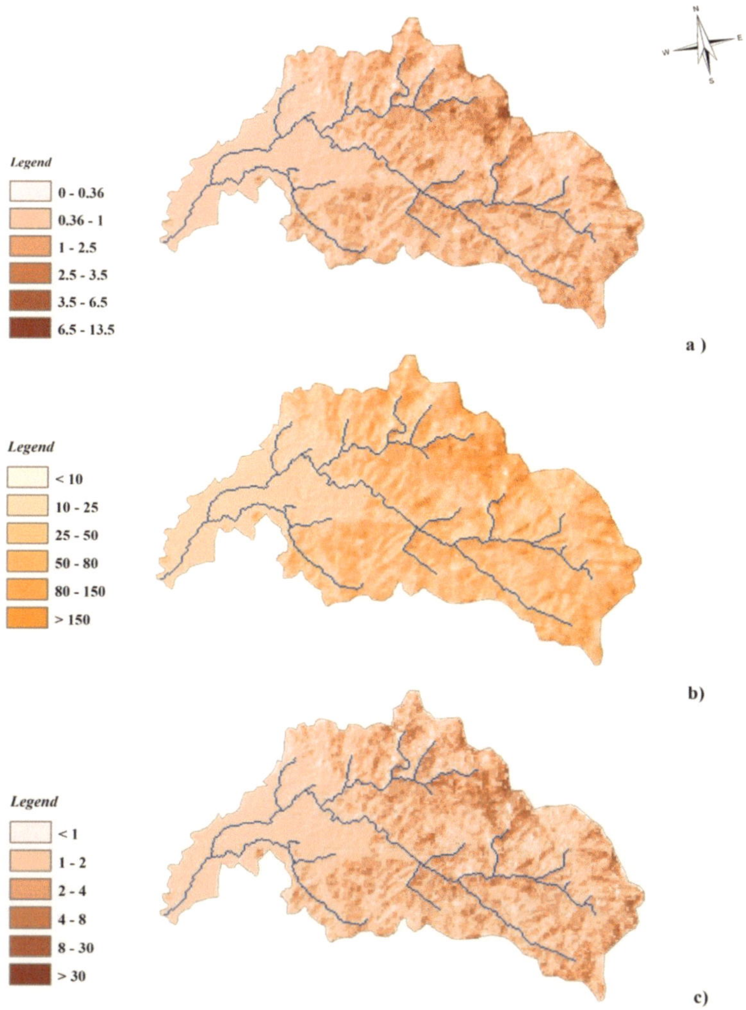

Figure 32.
Soil erosion potential (ton. ha^{-1} yr.$^{-1}$) with three scenarios of rainfall erosivity condition for the (a) mean value in dry season, (b) annual soil erosion, and (c) mean value of wet season.

predicted values from the RUSLE model are accurate and correlate well with prior research. Furthermore, the model effectively predicts soil loss potential for tropical catchments, reflecting the robustness of the prediction methodology used.

2.4.2 Discussion

In this research, annual and seasonal soil loss rates were generated for the Ulu Kinta Catchment, a tropical area. Various data sources were used to develop the SEPM model based on RUSLE input factors, which were stored as raster GIS layers in ArcGIS 10. The validation of the soil erosion potential model was conducted by examining each factor individually and comparing the results with previously published studies from areas with similar geo-environmental and rainfall characteristics. The annual average soil erosion potential was estimated at 34.72 tons ha^{-1} yr.$^{-1}$, while the

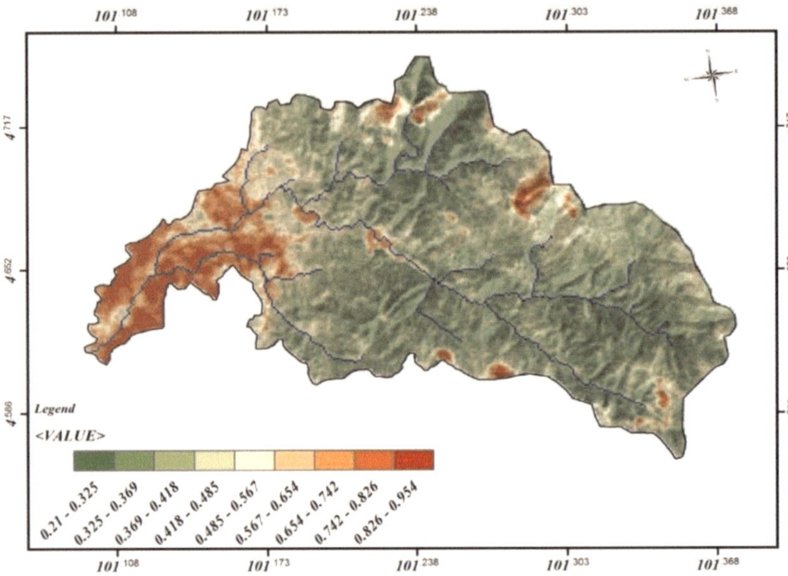

Figure 33.
Estimated cover management factor for the study area (CP factor).

		Numeric range (ton. ha^{-1} yr.$^{-1}$)					
Erosion risk level	**Intensity**	**Dry season**	**Distribution (%)**	**Wet season**	**Distribution (%)**	**Mean annual**	**Distribution (%)**
Very low	Slight	0–0.36	38.62	0–1	43.06	0–10	24.68
Low	Low	0.36–1	42.37	1–2	38.29	10–25	23.40
Moderate	Moderate	1–2.5	14.07	2–4	13.94	25–50	31.93
Severe	Severe	2.5–3.5	4.12	4–8	4.01	50–80	11.62
Very severe	Very severe	3.5–6.5	0.71	8–30	0.61	80–150	6.51
Extremely severe	Extremely severe	6.5–13.5	0.11	> 30	0.09	> 150	1.85

Table 12.
Derivation of the ordinal categories of soil erosion potential.

seasonal erosion potential ranged from 4.6 tons ha^{-1} yr.$^{-1}$ to 13.7 tons ha^{-1} yr.$^{-1}$. Additionally, the TRMM 3B43 dataset was utilized as an alternative source for generating rainfall erosivity, which is critical for further analysis.

Table 15 presents the classification of different soil erosion categories, indicating that nearly 80% of the study area falls under moderate to low erosion risk, while the remainder is classified as experiencing moderate to extremely severe erosion risk. The results were found to correlate well with similar studies conducted in other regions, aligning with the annual average soil loss rates reported. Furthermore, the sediment trap efficiency (STE) results showed a 71.9% correlation with the soil erosion potential model, indicating a strong and reasonable accuracy.

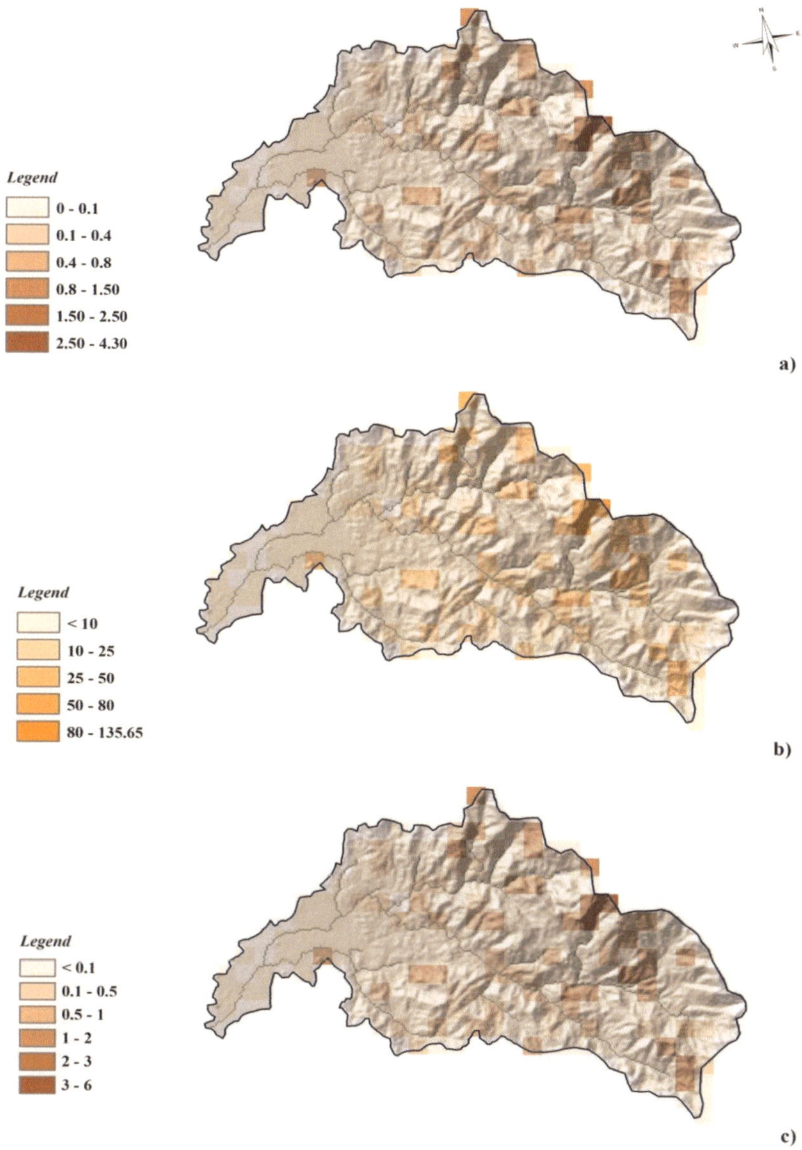

Figure 34.
Soil erosion potential (ton. ha^{-1} yr.$^{-1}$) on the basis of three scenarios of rainfall erosivity condition according to the TRMM dataset for the (a) mean value of soil erosion in dry season, (b) mean annual soil erosion, and (c) mean value of soil erosion in wet season.

3. Conclusion

The key findings from this study are as follows:

I. The soil erosion potential model (SEPM) effectively identifies hotspot erosion areas with high accuracy at the pixel scale, utilizing rainfall patterns, soil types, topography, and erosion control practices. This quantitative assessment provides average annual and seasonal soil loss estimates for the Ulu Kinta

		Numeric range (ton. ha^{-1} $yr.^{-1}$)					
Erosion risk level	**Intensity**	**Dry season**	**Distribution (%)**	**Wet season**	**Distribution (%)**	**Mean annual**	**Distribution (%)**
Very low	Slight	<0.1	69.87	<0.1	65.14	0–3	78.2
Low	Low	0.1–0.4	10.88	0.1–0.5	16.70	3–10	9.65
Moderate	Moderate	0.4–0.8	7.53	0.5–1	7.60	10–25	8.27
Severe	Severe	0.8–1.5	7.95	1–2	8.01	25–50	2.52
Very severe	Very severe	1.5–2.5	2.51	2–3	2.55	50–80	0.76
Extremely severe	Extremely severe	2.5–4.30	1.26	3–6	0	80–13	0.65

Table 13.
Derivation of the ordinal categories of soil erosion potential.

Area (references)	**Region**	**Method**	**SLR(ton h^{-1} $yr.^{-1}$)**		**Area (ha)**
			Range	**average**	
[68]	Taiwan	USLE	0 < A < 700	300	64,521
[69]	Penang	AGNPS	0 < A < 200	123	6309.25
[67]	Thailand	USLE	0 < A < 629	21.27	64628.9
[65]	Malaysia_Ulu Kinta	RUSLE	0 < A < 210	82.31	24,360
[64]	Malaysia_cameron highland	RUSLE	0 < A < 100	—	71,200
[70]	Kenyan highland	RUSLE	0 < A < 549	134	120
Present study	Malaysia_Ulu Kinta	RUSLE	0 < A < 780	34.72	30,752

Table 14.
Comparison between the current study and previous research in a similar area.

		Numeric range (ton. ha^{-1} $yr.^{-1}$)					
Erosion risk level	**Intensity**	**Dry season**	**Distribution (%)**	**Wet season**	**Distribution (%)**	**Mean annual**	**Distribution (%)**
Very low	Slight	<0.1	69.87	<0.1	65.14	0–3	78.15
Low	Low	0.1–0.4	10.88	0.1–0.5	16.70	3–10	9.65
Moderate	Moderate	0.4–0.8	7.53	0.5–1	7.60	10–25	8.27
Severe	Severe	0.8–1.5	7.95	1–2	8.01	25–50	2.52
Very severe	Very severe	1.5–2.5	2.51	2–3	2.55	50–80	0.76
Extremely severe	Extremely severe	2.5–4.30	1.26	3–6	0	80–13	0.65

Table 15.
Derivation of the ordinal categories of soil erosion potential seasonally and annually.

Catchment, integrating various data sources within a GIS framework, remote sensing techniques, and the well-established RUSLE equation. The average soil loss in the study area is 34.72 tons ha^{-1} $yr.^{-1}$, while TRMM-based erosivity yields a mean value of 7.9 tons ha^{-1} $yr.^{-1}$. The spatial distribution analysis highlights the significant roles of rainfall erosivity, soil erodibility, and the LS factor. Steep lands in the east and northeast, characterized by high rainfall and silt, are particularly vulnerable to erosion.

Overall, 19.98% of the area experiences severe to extremely severe erosion rates, leading to substantial soil depletion downstream. Additionally, the TRMM 3B43 dataset proves to be a valuable alternative for estimating annual SEPM, despite typically underestimating simulated soil loss rates compared to actual observations. This dataset serves as a crucial supporting resource for catchment management. Areas prone to higher erosion should be prioritized for soil erosion control measures. The model is thus a useful tool for assessing soil erosion risk in any catchment.

II. A significant percentage of the STE assessment procedure has shown in the annual amount of sediment yield with a value of 27.36 ton ha-1 yr-1. The model illustrated an appropriate application for sediment yield assessment.

III. A comparative analysis of published research revealed that the SEPM model for the Ulu Kinta Catchment, with a mean value of 34.72 tons ha^{-1} $yr.^{-1}$, aligns well with values from similar regions—ranging from a minimum annual average of 21.27 tons ha^{-1} $yr.^{-1}$ to a maximum of 300 tons ha^{-1} $yr.^{-1}$. This indicates that the predicted results fall within the expected range for tropical soil erosion.

Acknowledgements

The authors would like to thank the Perak Water Board (LAP) and Metropolitan Utilities Corporation (MUC) for their support and assistance in making this study a success. We also express our thanks to the respected reviewers for their valuable comments.

Author details

Mandana Abedini[1], Md Azlin Md Said[2] and Fauziah Ahmad[2*]

1 Abfan Consultant Eng, Tehran, Iran

2 Universiti Sains Malaysia, Nibong Tebal, Malaysia

*Address all correspondence to: cefahmad@usm.my

References

[1] Wischmeier W, Smith D. Predicting Rainfall Erosion Losses: A Guide to Conservation Planning [USA]. Agriculture handbook (USA): United States. Dept. of Agriculture; 1978

[2] Igwe PU, Onuigbo AA, Chinedu OC, Ezeaku II, Muoneke MM. Soil erosion: A review of models and applications. International Journal of Advanced Engineering Research and Science (IJAERS). 2017;**4**(12):138-150

[3] Morgan RPC. Soil Erosion and Conservation. John wiley & Sons; 2005

[4] Kinnell PIA. Event soil loss, runoff and the universal soil loss equation family of models: A review. Journal of Hydrology. 2010;**385**:384-397

[5] Lal R. Soil Erosion Research Methods. St. Lucie Press; 1994

[6] Zachar D. Soil Erosion. Elsevier Scientific Pub. Co; 1982

[7] Naik MG. Soil erosion and sediment yield modelling of watershed using fem, remote sensing and GIS. In: World Environmental and Water Resources Congress 2011. CA; 2011. pp. 1629-1640

[8] ESCM. Erosion and Sedimentation Control Measures (Escm) [Online]. Available from: http://dhn.iihr.uiowa.edu/runoff/erosion.htm

[9] Römkens MJM, Helming K, Prasad SN. Soil Erosion under Different Rainfall Intensities, Surface Roughness, and Soil Water Regimes. Catena. 2002;**46**: 103-123

[10] Zhang GH, Liu BY, Liu GB, He XW, Nearing MA. Detachment of undisturbed soil by shallow flow. Soil Science Society of America Journal. 2003;**67**:713-719

[11] Wang JG, Li ZX, Cai CF, Yang W, Ma RM, Zhang GB. Predicting physical equations of soil detachment by simulated concentrated flow in Ultisols (subtropical China). Earth Surface Processes and Landforms. 2012;**37**:633-641

[12] Kinnell PIA. Raindrop-impact-induced erosion processes and prediction: A review. Hydrological Processes. 2005;**19**:2815-2844

[13] Govers G. Empirical relationships for the transport capacity of overland flow. In: Erosion, Transport and Deposition Processes. IAHS Publication; 1990. pp. 45-63

[14] Op de Hipt F, Diekkrüger B, Steup G, Yira Y, Hoffmann T, Rodem M. Modeling the impact of climate change on water resources and soil erosion in a tropical catchment in Burkina Faso, West Africa. Catena. 2018;**163**:63-77

[15] Ergin M, Ediger V, Bodur MN, Okyar M. A review of modern sedimentation in the Golden Horn estuary (sea of Marmara), Turkey. Bollettino D Oceanolog a Teorica Ed Applicata, III. 1990;**2**:135-151

[16] Lewis DW, McConchie DM. Practical Sedimentology. 2nd ed. NY (United States), Chapman Hall: New York; 1994

[17] Abedini M, Said M, Md A, Ahmad F. Effectiveness of check dam to control soil erosion in a tropical catchment (the ulu Kinta Basin). Catena. 2012;**97**:63-70

[18] Xiong M, Sun R, Chen L. Effects of soil conservation techniques on water

erosion control: A global analysis. Science of the Total Environment. 2018; **645**:753-760

[19] Vaezi AR, Abbasi M, Keesstra S, Cerdà A. Assessment of soil particle erodibility and sediment trapping using check dams in small semi-arid catchments. Catena. 2017;**157**:63-77

[20] Merritt W, Letcher R, Jakeman A. A review of erosion and sediment transport models. Environmental Modelling & Software. 2003;**18**:761-799

[21] Meyer LD, Wischmeier W. Mathematical simulation of the process of soil erosion by water. American Society of Agricultural Engineers. 1969; **12**:754-758

[22] Lane L, Shirley E, Singh V. Modelling erosion on hillslopes. In: Anderson MG, editor. Modelling Geomorphological Systems. John Wiley & Sons Ltd; 1988

[23] Wahla SS, Kazmi JH, Tariq A. Mapping and monitoring of spatio-temporal land use and land cover changes and relationship with normalized satellite indices and driving factors. In: Geology, Ecology and Landscapes. Taylor & Francis; 2023

[24] Abedini M, Said M, Md A, Ahmad F. Integration of statistical and spatial methods for distributing precipitation in tropical areas. Hydrology Research. 2013;**44**(6):982-994

[25] Lal R. Soil degradation by erosion. Land Degradation & Development. 2001;**12**:519-539

[26] Smith RE, Quinton J, Goodrich DC, Nearing M. Soil-erosion models: Where do we really stand? Earth Surface Processes and Landforms. 2010;**35**: 1344-1348

[27] Mehta D, Hadvani J, Kanthariya D, Sonawala P. Effect of land use land cover change on runoff characteristics using curve number: A GIS and remote sensing approach. International Journal of Hydrology Science and Technology. 2023;**16**(1, 1):-16

[28] Park S, Oh C, Jeon S, Jung H, Choi C. Soil erosion risk in Korean watersheds, assessed using the revised universal soil loss equation. Journal of Hydrology. 2011;**399**:263-273

[29] Ahmad NSBN, Mustafa FB, Yusoff SYM, Didams G. A systematic review of soil erosion control practices on the agricultural land in Asia. International Soil and Water Conservation Research. 2020;**8**(2):103-115 28

[30] Sakuno NRR, Guiçardi ACF, Spalevic V, Avanzi JC, Silva MLN, Mincato RL. Adaptation and application of the erosion potential method for tropical soils. Revista Ciência Agronômica. 2020;**51**(1):e20186545

[31] Nearing MA, Jetten V, Baffaut C, Cerdan O, Couturier A, Hernandez M, et al. Modeling response of soil erosion and runoff to changes in precipitation and cover. Catena. 2005;**61**:131-154

[32] Galetovic JR, Toy TJ, Foster GR. Guidelines for the Use of the Revised Universal Soil Loss Equation (RUSLE), Version 1.06, on Mined Lands, Construction Sites and Reclaimed Lands. Office of Technical Transfer, Western Regional Coordination Center, Office of Surface Mining. 1998

[33] Lu D, Li G, Valladares GS, Batistella M. Mapping soil erosion risk in Rondônia, Brazilian Amazonia: Using RUSLE, remote sensing and GIS. Land Degradation & Development. 2004;**15**: 499-512

[34] Renard KG, Freimund JR. Using monthly precipitation data to estimate the R-factor in the revised USLE. Journal of Hydrology. 1994a;**157**:287-306

[35] Millward AA, Mersey JE. Adapting the RUSLE to model soil erosion potential in a mountainous tropical watershed. Catena. 1999;**38**:109-129

[36] Tiwari A, Risse L, Nearing M. Evaluation of WEPP and its comparison with USLE and RUSLE. ASAE American Society of Agricultural. 2000;**43**(PART 5):1129-1136

[37] Pham TN, Yang D, Kanae S, Oki T, Musiake K. Application of RUSLE model on global soil erosion estimate. Annual Journal of Hydraulic Engineering, JSCE. 2001;**45**:811-816

[38] Walling. Modelling Erosion, Sediment Transport and Sediment Yield. International Hydrological Programme: UNESCO; 2002

[39] USDA-ARS. Draft Scientific Documentation Revised Universal Soil Loss Equation, Version 2; 2008

[40] Foster GR, Toy TJ, Renard KG. Comparision of the USLE, RUSLE1.06c, and RUSLE2 for application to highly distributed lands. In: First Interagency Conference on Research in the Watersheds. Benson; 2003

[41] USDA-ARS. Draft Scientific Documentation Revised Universal Soil Loss Equation, Version 2; 2005

[42] Tew KH, Roslan ZA. Soil erosion and sedimentation assessment, control and management plan for the tropical region. In: Tropical Residual Soils Engineering. London: Taylor & Francis; 2004

[43] Dabney SM, Yoder DC, Vieira DAN, Bingner RL. Enhancing RUSLE to include runoff-driven phenomena. Hydrological Processes. 2011;**25**: 1373-1390

[44] Rajbanshi J, Das S, Paul R. Quantification of the effects of conservation practices on surface runoff and soil erosion in croplands and their trade-off: A meta-analysis. Science of the Total Environment. 2023;**864**(4):161015

[45] JICA. The study on the establishment of the River Basin information system in Malaysia. In: SSS. CR3, editor. Final Report. Japan International Cooperation Agency; 1999

[46] Karaburun A. Estimation of C factor for soil erosion modeling using NDVI in Buyukcekmece watershed. Ozean Journal of Applied Sciences. 2010;**3**:77-85

[47] Gelfand AE, Banerjee S, Gamerman D. Spatial process modelling for univariate and multivariate dynamic spatial data. Environmetrics. 2005;**16**: 465-479

[48] Islam MN, Uyeda H. Use of Trmm in determining the climatic characteristics of rainfall over Bangladesh. Remote Sensing of Environment. 2007;**108**:264-276

[49] Abedini M, Md Said MA, Ahmad F. Clustering approach on land use land cover classification of landsat TM over Ulu Kinta catchment. World Applied Sciences Journal. 2012;**17**:809-817

[50] EI-Swaify S, Dangler E, Armstrong C. Soil erosion by water in tropics. University of Hawaii, College of Agriculture, Research Extension Series. 1982;**24**:173

[51] Immerzeel WW, Rutten MM, Droogers P. Spatial downscaling of TRMM precipitation using vegetative response on the Iberian Peninsula.

Remote Sensing of Environment. 2009; **113**:362-370

[52] Zhang H, Loáiciga HA, Ha D, Qingyun D. Spatial and temporal downscaling of TRMM precipitation with novel algorithms. Journal of Hydrometeology. 2020;**21**(6):1259-1278

[53] NASA. Tropical Rainfall Measuring Mission (Trmm) [Online]. 2013. Available from: http://trmm.gsfc.nasa.gov/

[54] Saha S, Gayen A, Pourghasemi HR, Tiefenbacher JP. Identification of soil erosion-susceptible areas using fuzzy logic and analytical hierarchy process modeling in an agricultural watershed of Burdwan district, India. Environmental Earth Science. 2019;**78**(649)

[55] ESRI. 1995-2010. ArcGIS Resource Center. [Online]. Esri

[56] Tew KH. Production of Malaysian Soil Erodibility Nomograph in Relation to Soil Erosion Issues. Tew kia Hui; 1999

[57] Sorooshian S, Gao X, Hsu K, Maddox RA, Hong Y, Gupta HV, et al. Diurnal variability of tropical rainfall retrieved from combined goes and Trmm satellite information. Journal of Climate. 2002;**15**: 983-1001

[58] Albert JM. Hydraulic Analysis and Double Mass Curves of the Middle Rio Grande from Cochiti to San Marcial. New Mexico: MASTER OF SCIENCE, Colorado State University; 2004

[59] Verlinde J. TRMM rainfall downscaling in the Pangani Basin in Tanzania. Master of Science, Delft University of Technology; 2011

[60] Huffman GJ, Adler RF, Arkin P, Chang A, Ferraro R, Gruber A, et al. The global precipitation climatology project (Gpcp) combined precipitation dataset. Bulletin of the American Meteorological Society. 1997;**78**:5-20

[61] Searcy JK, Hardison CH. Double-Mass Curves, Manual of Hydrology. U.S. Geological Survey Water-Supply. USGS: Washington; 1960

[62] Yu B, Hashim GM, Eusof Z. Estimating the R-factor with limited rainfall data: A case study from Peninsular Malaysia. Journal of soil and Water Conservation. 2001;**56**: 101-105

[63] Shamshad A, Azhari MN, Isa MH, Hussin WMAW, Parida BP. Development of an appropriate procedure for estimation of RUSLE Ei30 index and preparation of Erosivity maps for Pulau Penang in Peninsular Malaysia. CATENA. 2008a;**72**:423-432

[64] Teh SH. Soil Erosion Modeling using RUSLE and GIS on Cameron Highlands. Malaysia for Hydropower Developement. Master Degree, University of Iceland & University of Akureyri; 2011

[65] Bawahidi KSY. Integrated Land Use Change Analysis for Soil Erosion Study in Ulu Kinta Catchment. Ph.D, University Sains Malaysia; 2006

[66] DID. Guidline for Erosion and Sediment Control in Malaysial. Departmnet of Irregration and Drainage Malaysia (DID): Department of Irreggration and Drainage Ministry of Agriculture; 2010

[67] Krishna Bahadur K. Mapping soil erosion susceptibility using remote sensing and GIS: A case of the Upper Nam Wa Watershed, Nan Province, Thailand. Environmental Geology. 2009; **57**:695-705

[68] Lin CY, Lin WT, Chou WC. Soil erosion prediction and sediment yield estimation: The Taiwan experience. Soil and Tillage Research. 2002;**68**:143-152

[69] Shamshad A, Leow CS, Ramlah A, Wan Hussin WMA, Mohd. Sanusi SA. Applications of annagnps model for soil loss estimation and nutrient loading for Malaysian conditions. International Journal of Applied Earth Observation and Geoinformation. 2008b;**10**:239-252

[70] Angima SD, Stott DE, O'Neill MK, Ong CK, Weesies GA. Soil erosion prediction using RUSLE for Central Kenyan highland conditions. Agriculture, Ecosystems & Environment. 2003;**97**:295-308

Chapter 5

Perspective Chapter: Big Data and Deep Learning in Hydrological Modeling

Li Zhou

Abstract

This chapter delves into the integration of physical mechanisms with deep learning models to enhance the interpretability and accuracy of hydrological process modeling. In the era of big data and rapid advancements in AI, the synergy between traditional hydrological principles and machine learning opens new opportunities for improved water resource management, flood prediction, and drought monitoring. The chapter presents a comprehensive framework that leverages vast datasets from sources such as remote sensing, reanalysis data, and in situ monitoring. It explores the potential of deep learning models, particularly when combined with physical insights, to address challenges in data-scarce regions, improving the accuracy and transparency of predictions. By analyzing the strengths and limitations of current approaches, the study highlights the value of hybrid models in balancing accuracy and interpretability. These models not only enhance predictive performance but also provide more transparent insights into the underlying hydrological processes. This integration contributes to sustainable water management, disaster resilience, and climate adaptation, pushing forward both scientific progress and practical applications. The chapter offers valuable methodologies and case studies that underscore the importance of domain knowledge in the development of explainable and reliable predictive models, reshaping the future of hydrological forecasting.

Keywords: hydrological modeling, big data analytics, data-driven hydrology, runoff simulation, water management

1. Introduction

As societal development progresses alongside the escalating challenges posed by climate change, the management of water resources has become increasingly critical. Effective water resource management is essential for sustaining ecosystems and supporting human activities, particularly in regions vulnerable to flooding and drought. Accurate runoff simulation and prediction play pivotal roles in various applications, including water resource planning, flood and drought early warning systems, and watershed evaluation [1].

The rainfall-runoff process is influenced by numerous complex physical mechanisms [2], characterized by significant spatial and temporal uncertainties. Traditional hydrological models, such as the SWAT (Soil and Water Assessment Tool) and HEC-HMS (Hydrologic Engineering Center's Hydrologic Modeling System), rely on detailed land use, soil, and topographic data to simulate hydrological processes. While these models have been instrumental in advancing hydrology, they often require extensive calibration and data inputs, which can be a significant limitation in ungauged basins—regions that lack sufficient hydrological data for model calibration.

In recent years, the advent of big data and advancements in artificial intelligence have spurred interest in data-driven approaches, particularly deep learning methods. Deep learning models, such as long short-term memory (LSTM) networks and convolutional neural networks (CNNs), excel at identifying patterns in large datasets without the need for explicitly modeling complex physical processes. These models have demonstrated significant promise in hydrological forecasting, particularly in ungauged basins where traditional models struggle due to a lack of observational data. However, while deep learning models offer improved accuracy, they are often criticized for their "black box" nature, lacking transparency in their internal workings. This raises important questions about the interpretability of these models in hydrological contexts, where understanding the mechanisms behind predictions is crucial for both scientific inquiry and practical applications.

This chapter presents the research progress and future directions in hydrological modeling using big data and deep learning. It begins by discussing the growing need for more accurate hydrological models, especially in data-scarce regions such as ungauged basins, where traditional models struggle. The chapter then delves into the limitations of conventional hydrological models, such as their reliance on extensive physical data and the challenges of calibration in these regions. It highlights the advancements made through deep learning models, particularly long short-term memory (LSTM) networks, which can identify patterns in large datasets without the need for explicit physical modeling. The integration of physical hydrological processes with deep learning models is emphasized as a key development, enhancing both accuracy and interpretability. Furthermore, the chapter explores the use of hybrid models that combine deep learning with traditional hydrological approaches, offering a balanced solution to the challenges of runoff prediction. The chapter concludes by discussing future research directions, including the need for further exploration of physically interpretable deep learning models and their application in hydrological forecasting and water resource management.

2. Runoff prediction in ungauged basins

At present, runoff simulation and prediction in data-scarce or ungauged basins (hereafter referred to as "ungauged") is a particularly challenging issue faced by the international hydrological community [3, 4]. Globally, there are numerous watersheds that lack ground observation data, and these basins are often areas where hydrological disasters, such as floods, debris flows, and landslides, occur frequently [5, 6]. The lack of reliable data in these areas complicates not only the understanding of hydrological processes but also the implementation of effective management and mitigation strategies. Consequently, enhancing research on runoff prediction in ungauged basins has emerged as a crucial and urgent priority in the field of hydrology.

The term "ungauged" specifically refers to basins that either lack historical hydrological data entirely or have insufficient measurements to inform predictive modeling (in this project, it particularly denotes basins without measured runoff data). This gap in data presents a significant barrier to the accurate forecasting of runoff, which is essential for water resource management, disaster risk reduction, and environmental sustainability. In response to these challenges, the International Association of Hydrological Sciences (IAHS) initiated a 10-year scientific program in 2003, known as PUB (Prediction in Ungauged Basins) [7]. This program aimed to systematically address the complexities of runoff prediction in ungauged basins and has since catalyzed considerable research efforts.

Over the past two decades, scholars have made substantial progress by applying various process-based hydrological models, including conceptual models, physical models, as well as lumped and distributed models. These efforts have yielded advancements in several key areas, such as regionalization, parameter transfer, watershed similarity analysis, and donor-basin methods [8, 9]. Despite these achievements, significant challenges remain. The high demand for comprehensive data continues to be a major limitation, as does the substantial computational cost associated with sophisticated modeling techniques. Furthermore, the accuracy of runoff predictions in ungauged basins still falls short compared to results derived from calibrated models in individual basins [10, 11]. This discrepancy underscores the need for innovative approaches that can leverage available data more effectively, potentially incorporating machine learning techniques and integrating hydrological understanding into model frameworks.

3. Data-driven models based on deep learning

In 2013, the International Association of Hydrological Sciences (IAHS) launched a research initiative known as "Panta Rhei," aimed at better explaining and addressing global hydrological and water resource challenges by integrating big data information and empirical knowledge. This initiative has significantly promoted the application of data-driven models in the field of hydrology [12]. However, there has been ongoing debate regarding the relative advantages of data-driven models versus process-driven models. With the advent of the era of big data and artificial intelligence, deep learning models have demonstrated outstanding performance in various fields [13, 14]. In the realm of hydrology and water resources, traditional process-driven hydrological models face numerous challenges in terms of accuracy and efficiency due to the requirement for extensive empirical parameters and precise descriptions of complex physical processes [15]. In contrast, deep learning models can automatically identify patterns directly from large datasets without the need for explicitly modeling physical processes, making them highly promising for runoff prediction in ungauged basins [16].

However, deep learning models have often been criticized for their "black box" nature, as they lack transparency in explaining their internal workings. This lack of physical interpretability is particularly problematic in hydrology, where understanding the internal mechanisms of rainfall-runoff processes and the characteristics of the underlying surface is crucial for scientific understanding and practical application [15, 17]. The key scientific challenge in this field is how to imbue deep learning models with physical interpretability.

The emerging research direction in hydrology emphasizes the development of deep learning models that integrate the physical characteristics of hydrological basins, particularly in ungauged or data-scarce regions, to enhance the accuracy of runoff predictions [18]. Traditional hydrological models, which rely heavily on physical laws and processes, often struggle in such conditions due to the lack of observational data [19]. In contrast, deep learning models have shown considerable potential by learning patterns directly from data, which can be particularly advantageous in these contexts [20].

However, these models are frequently criticized for their "black-box" nature, making it difficult to interpret how they arrive at predictions. **Table 1** shows the pros and cons of different models. A key focus of current research is not only to improve the accuracy of deep learning models in runoff forecasting but also to enhance their interpretability by incorporating hydrological processes and physical characteristics into the model's architecture [21]. By embedding the knowledge of rainfall-runoff dynamics and basin characteristics into deep learning models, researchers aim to bridge the gap between data-driven approaches and process-based hydrological understanding [22]. This integration could offer more reliable and physically consistent predictions, especially in ungauged basins where traditional models are less effective.

In addition to improving prediction accuracy, this research seeks to explore the internal mechanisms of deep learning models—how they use hydrological variables to reflect real-world processes. The goal is to uncover and explain the relationships between model structures, parameters, and hydrological processes. By doing so, researchers can develop more transparent and interpretable models, enhancing both the trustworthiness and applicability of deep learning in hydrological forecasting.

This area of study represents a new frontier in hydrology and water resources research, with significant implications for both theory and practice. The integration of physically interpretable deep learning models could not only improve predictive performance in ungauged basins but also deepen our understanding of hydrological processes, potentially leading to breakthroughs in flood forecasting, drought management, and water resource planning. As such, this interdisciplinary approach is gaining attention as a major focus in the field, offering the potential for significant advancements in scientific knowledge and practical water management solutions.

Model type	Advantages	Disadvantages
Traditional hydrological models (e.g., SWAT, VIC)	Based on physical processes Well-established methodologies Can incorporate climate and land-use changes	Requires detailed input data May not perform well in data-scarce areas High computational cost
Machine learning models (e.g., LSTM, Random Forest)	Learns from historical data Works well with sparse and incomplete data Fast computation	Lacks interpretability Requires large amounts of training data May not generalize well outside of training data
Hybrid models (e.g., physically-based + AI models)	Combines the strengths of both approaches Improved accuracy Can model complex interactions	Difficult to design and calibrate Computational complexity can be high

Table 1.
Comparison of model advantages and disadvantages.

4. Research progress of deep learning in runoff prediction for ungauged basins

In recent years, deep learning has emerged as a powerful artificial intelligence technology, rapidly developing and making significant inroads across various domains. Particularly in hydrological forecasting, especially for runoff prediction, deep learning models have showcased substantial potential [23, 24]. Among the prevalent deep learning models employed in this field are feedforward neural networks, convolutional neural networks (CNNs), and recurrent neural networks (RNNs). Notably, LSTM (basic structure shown in **Figure 1**), an advanced variant of RNN, effectively mitigates the challenges of gradient vanishing and gradient explosion commonly encountered in traditional RNNs. By incorporating specialized memory cells, LSTM markedly enhances the model's capacity to retain and leverage historical information, solidifying its status as the most widely utilized deep learning architecture in contemporary hydrological simulations [25]. The application of LSTM in rainfall-runoff simulations has witnessed exponential growth, reflecting its effectiveness in managing temporal dependencies inherent in hydrological data.

In the realm of deep learning, runoff prediction for ungauged basins represents a critical challenge, where the generalization capabilities of pre-trained neural networks are harnessed to forecast new data samples that were not included in the original training dataset [26]. Studies have highlighted LSTM's considerable potential in this context, showcasing its effectiveness in runoff prediction for ungauged basins [9, 27]. For instance, Kratzert et al. [14] applied LSTM to runoff forecasting, and later, they used the large-sample hydrological dataset CAMELS (Catchment Attributes and Meteorology for Large-sample Studies) to conduct cross-experiments

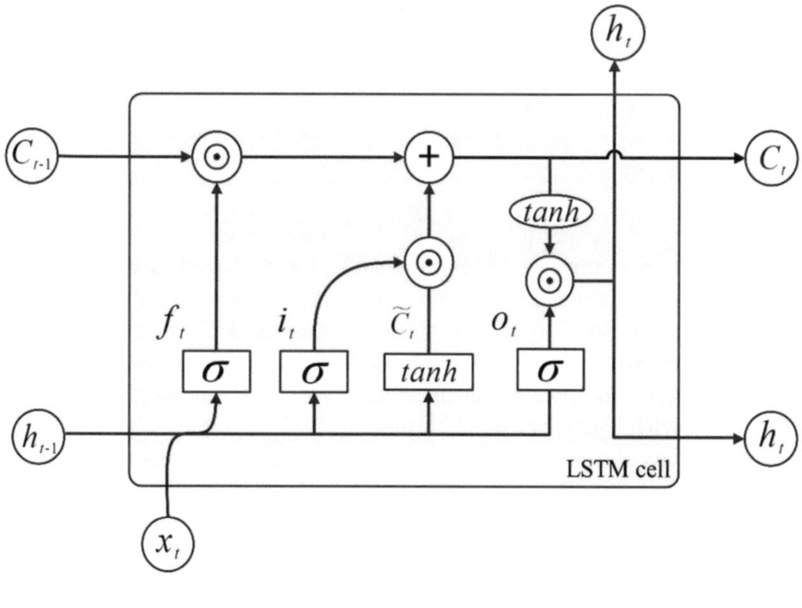

Figure 1.
Basic structure of LSTM cell. Note: σ and tanh *represent sigmoid and hyperbolic tangent activation function, respectively.* ⊙ *is Hadamard (elementwise) product operator.* ft, it, *and* Ot *represent forget gate, update gate, and output gate, respectively.* xt *denotes the input data,* Ct *denotes the cell state,* ht *denotes the hidden state, and* $\widetilde{C}_t$ *denotes the candidate memory cell state.*

in 531 basins, achieving better runoff prediction results than traditional hydrological models [28]. Hashemi et al. [13] studied the impact of regional hydrological information on LSTM model training and hyperparameter optimization across 361 basins in France. These studies suggest that for LSTM applications in runoff prediction for ungauged basins, large-sample hydrological datasets must be employed to allow the model to learn a diverse array of hydrological processes, rather than being trained on just a few basins [11, 14, 28, 29].

However, the use of LSTM in runoff prediction for ungauged basins also faces many challenges. First, parameter tuning and model training for LSTM become more complex in ungauged scenarios, necessitating the development of more effective methods and strategies. Second, there is an urgent need for research into how to build regionalized LSTM models that incorporate the physical hydrological characteristics of basins, how hyperparameter selection affects runoff prediction accuracy, and how to interpret the hydrological behaviors of rainfall-runoff processes within deep learning models. Addressing these challenges will improve the accuracy of runoff prediction under ungauged conditions and advance the application of deep learning in hydrology.

Furthermore, the integration of domain knowledge and physical hydrological principles into deep learning models presents a promising avenue for future research. By combining traditional hydrological modeling techniques with deep learning approaches, researchers can develop hybrid models that leverage the strengths of both methodologies. This integrative approach could enhance model interpretability and robustness, facilitating better understanding and prediction of hydrological processes. As research progresses, the focus on developing standardized frameworks for data collection, model training, and evaluation will be essential in fostering the wider adoption of deep learning techniques in runoff prediction for ungauged basins. Ultimately, overcoming the existing challenges will not only enhance the accuracy and reliability of hydrological forecasts but also contribute to improved water resource management in regions lacking comprehensive gauging networks.

5. Research progress on the physical interpretability of deep learning in runoff prediction

Despite the powerful capabilities of deep learning models in runoff prediction, their "black box" nature hinders widespread adoption and application. Compared to traditional hydrological models, deep learning models lack interpretability, making it difficult to understand the logical relationships behind their predictions, which reduces trust in the results. Current deep learning-based runoff prediction models primarily focus on improving simulation accuracy but often neglect the integration of model structures with hydrological principles.

Traditional physical models are based on the physical mechanisms of hydrological factors and usually require detailed information about the underlying surface. For example, the SWAT model [30] incorporates detailed land use, soil, and topographic information to simulate runoff generation and transport processes. However, they have disadvantages such as long computation times and high costs. Data-driven models, in contrast, do not require knowledge of the formation processes of hydrological factors, are easy to construct, and possess strong nonlinear fitting capabilities. Nonetheless, shallow data-driven models based on machine learning, such as artificial neural networks (ANNs) or support vector machines (SVMs), cannot effectively transmit historical information and have limited learning capacity, especially for multi-step outputs [31]. Thus, utilizing advanced deep learning methods, such as long short-term memory (LSTM) networks, for the simulation and prediction of hydrological factors presents an excellent alternative [14, 25]. Existing studies predominantly adopt a single LSTM model, without considering the interrelationships among input data or the differing importance of historical data at different time steps.

To date, research on the interpretability of deep learning models in hydrological predictions, especially in the context of rainfall-runoff processes, remains limited [22]. This is a crucial area of study as traditional hydrological models are deeply rooted in physical principles, while deep learning models are often seen as "black-box" approaches due to their lack of transparency. Recent studies have categorized interpretability approaches into two main types: integrating deep learning with physical mechanism models and using physical principles to guide the development of deep learning models [32]. For example, Xie et al. [33] used a hybrid approach where physically-based loss functions were introduced to guide the training of deep learning models, which significantly improved the model's interpretability in terms of runoff generation processes. Similarly, Nimai et al. [34] explored the integration of deep learning models with conceptual hydrological models, showing that this hybrid approach can better capture the spatial variability of hydrological processes while retaining some level of interpretability.

The first approach involves replacing certain components of traditional hydrological models with deep learning models, such as training hydrological model parameters using neural networks. For example, Kratzert et al. [28] successfully applied LSTM networks to predict streamflow in ungauged basins, showing that deep learning models can outperform traditional hydrological models by capturing complex temporal dependencies in data. This approach demonstrates that, although deep learning models may not directly model physical processes, they can still be highly effective in scenarios where traditional models struggle due to data scarcity or complexity.

The second approach involves guiding the structure and training of deep learning models with physical insights. Sun et al. [35] introduced physics-guided recurrent neural networks for global hydrological modeling, where physical laws were embedded into the model's loss function to ensure that predictions remained consistent with known hydrological principles. This type of model represents a more transparent and interpretable form of deep learning because it incorporates physical constraints, reducing the risk of generating physically implausible predictions.

Despite these advancements, significant challenges remain. One key issue is how to fully integrate hydrological process knowledge into deep learning architectures in a way that balances physical interpretability and prediction accuracy. For instance, Jiang et al. [36] highlighted the importance of making machine learning models not only accurate but also physically interpretable, particularly for applications in high-resolution hydrological simulations.

Going forward, a major research direction will be the development of hybrid models that can effectively combine the strength of data-driven approaches with the explanatory power of physical models. By embedding hydrological concepts into the deep learning frameworks—whether through guiding model architecture, feature engineering, or designing loss functions—it is possible to improve both model transparency and performance. Exploring these hybrid approaches is a key challenge in the field of hydrology, with significant potential for improving predictions in both gauged and ungauged basins.

It is evident that the integration of deep learning and physical mechanisms is often loosely coupled, mainly through modifications to the loss function. However, further exploration is needed to determine whether there are deep learning models more suited for runoff prediction, how hydrological physical mechanisms can be more thoroughly integrated into deep learning model structures, and how to build models that are both physically consistent and have strong predictive performance, thereby driving the discovery of more hydrological patterns. Future work should aim to explore the connections between deep learning model structures and parameters with hydrological physical variables, establish correlations between model internal units and hydrological concepts, and develop interpretable deep learning models for runoff prediction. Enhancing the transparency and trustworthiness of these models remains a frontier topic in current research on deep learning in runoff prediction.

6. Conclusion

In conclusion, this chapter underscores the critical need for integrating physical mechanisms into deep learning models within hydrological process modeling. By bridging the gap between traditional hydrological principles and modern machine learning techniques, we can enhance both the interpretability and accuracy of predictive models. The exploration of hybrid frameworks not only provides insights into model behavior but also reinforces the applicability of these models in real-world scenarios, especially in data-scarce regions. This approach fosters greater trust and usability, essential for effective water resource management. Ultimately, the findings advocate for a paradigm shift towards collaborative modeling efforts that prioritize transparency, reliability, and the incorporation of domain expertise, thereby advancing the field of hydrology and its applications in addressing pressing water-related challenges.

Acknowledgements

We gratefully acknowledge the Fundamental Research Funds for the Central Universities and the Natural Science Foundation Youth Project (2024NSFSC0984) from Science and Technology Department of Sichuan Province.

Author details

Li Zhou
Institute for Disaster Management and Reconstruction, Sichuan University-Hong Kong Polytechnic University, Chengdu, China

*Address all correspondence to: zhouli.scu@gmail.com

References

[1] Dietze M, Ozturk U. A flood of disaster response challenges. Science. 2021;**373**:1317-1318. DOI: 10.1126/science.abm0617

[2] Zhou L, Koike T, Takeuchi K, et al. A study on availability of ground observations and its impacts on bias correction of satellite precipitation products and hydrologic simulation efficiency. Journal of Hydrology. 2022;**610**:127595. DOI: 10.1016/j.jhydrol.2022.127595

[3] Salinas JL, Laaha G, Rogger M, et al. Comparative assessment of predictions in ungauged basins - part 2: Flood and low flow studies. Hydrology and Earth System Sciences. 2013;**17**:2637-2652. DOI: 10.5194/hess-17-2637-2013

[4] Hrachowitz M, Savenije HHG, Blöschl G, et al. A decade of predictions in ungauged basins (PUB)—A review. Hydrological Sciences Journal. 2013;**58**:1198-1255. DOI: 10.1080/02626667.2013.803183

[5] Xu Y, Lin K, Hu C, et al. Deep transfer learning based on transformer for flood forecasting in data-sparse basins. Journal of Hydrology. 2023;**625**:129956. DOI: 10.1016/j.jhydrol.2023.129956

[6] Jhong Y, Chen C, Jhong B, et al. Optimization of LSTM parameters for flash flood forecasting using genetic algorithm. Water Resources Management. 2024;**38**:1141-1164. DOI: 10.1007/s11269-023-03713-8

[7] Sivapalan M, Takeuchi K, Franks SW, et al. IAHS decade on predictions in ungauged basins (PUB), 2003-2012: Shaping an exciting future for the hydrological sciences. Hydrological Sciences Journal. 2003;**48**:857-880. DOI: 10.1623/hysj.48.6.857.51421

[8] Guo J, Liu Y, Zou Q, et al. Study on optimization and combination strategy of multiple daily runoff prediction models coupled with physical mechanism and LSTM. Journal of Hydrology. 2023;**624**:129969. DOI: 10.1016/j.jhydrol.2023.129969

[9] Ma K, He D, Liu S, et al. Novel time-lag informed deep learning framework for enhanced streamflow prediction and flood early warning in large-scale catchments. Journal of Hydrology. 2024;**631**:130841. DOI: 10.1016/j.jhydrol.2024.130841

[10] Hashemi R, Javelle P, Delestre O, et al. Closing the data gap: Runoff prediction in fully ungauged settings using LSTM. Hydrology and Earth System Sciences Discussions. 2023;**2023**:1-41. DOI: 10.5194/hess-2023-282

[11] Kratzert F, Gauch M, Klotz D, et al. HESS opinions: Never train an LSTM on a single basin. Hydrology and Earth System Sciences Discussions. 2024;**2024**:1-19. DOI: 10.5194/hess-2023-275

[12] Montanari A, Young G, Savenije HHG, et al. "Panta Rhei—Everything flows": Change in hydrology and society—The IAHS scientific decade 2013-2022. Hydrological Sciences Journal. 2013;**58**:1256-1275. DOI: 10.1080/02626667.2013.809088

[13] Hashemi R, Brigode P, Garambois P, et al. How can we benefit from regime information to make more effective use of long short-term memory (LSTM) runoff models? Hydrology and Earth System Sciences. 2022;**26**:5793-5816. DOI: 10.5194/hess-26-5793-2022

[14] Kratzert F, Klotz D, Brenner C, et al. Rainfall–runoff modelling using long short-term memory (LSTM) networks. Hydrology and Earth System Sciences. 2018;**22**:6005-6022. DOI: 10.5194/hess-22-6005-2018

[15] Arsenault R, Martel J, Brunet F, et al. Continuous streamflow prediction in ungauged basins: Long short-term memory neural networks clearly outperform traditional hydrological models. Hydrology and Earth System Sciences. 2023;**27**:139-157. DOI: 10.5194/hess-27-139-2023

[16] Wi S, Steinschneider S. On the need for physical constraints in deep learning rainfall–runoff projections under climate change: A sensitivity analysis to warming and shifts in potential evapotranspiration. Hydrology and Earth System Sciences. 2024;**28**:479-503. DOI: 10.5194/hess-28-479-2024

[17] Feng D, Beck H, Lawson K, et al. The suitability of differentiable, physics-informed machine learning hydrologic models for ungauged regions and climate change impact assessment. Hydrology and Earth System Sciences. 2023;**27**:2357-2373. DOI: 10.5194/hess-27-2357-2023

[18] Hasan F, Medley P, Drake J, et al. Advancing hydrology through machine learning: Insights, challenges, and future directions using the CAMELS, caravan, GRDC, CHIRPS, PERSIANN, NLDAS, GLDAS, and GRACE datasets. Water. 2024;**16**:1904

[19] Liu L, Zhou L, Gusyev M, et al. Unravelling and improving the potential of global discharge reanalysis dataset in streamflow estimation in ungauged basins. Journal of Cleaner Production. 2023;**419**:138282. DOI: 10.1016/j.jclepro.2023.138282

[20] Mount NJ, Maier HR, Toth E, et al. Data-driven modelling approaches for socio-hydrology: Opportunities and challenges within the Panta Rhei science plan. Hydrological Sciences Journal. 2016;**61**:1192-1208. DOI: 10.1080/02626667.2016.1159683

[21] Kim J, Han H, Johnson LE, et al. Hybrid machine learning framework for hydrological assessment. Journal of Hydrology. 2019;**577**:123913. DOI: 10.1016/j.jhydrol.2019.123913

[22] Yue J, Zhou L, Du J, et al. Runoff simulation in data-scarce alpine regions: Comparative analysis based on LSTM and physically based models. Water. 2024;**16**:2161

[23] Tsai W, Feng D, Pan M, et al. From calibration to parameter learning: Harnessing the scaling effects of big data in geoscientific modeling. Nature Communications. 2021;**12**:5988. DOI: 10.1038/s41467-021-26107-z

[24] van Natijne AL, Lindenbergh RC, Bogaard TA. Machine learning: New potential for local and regional deep-seated landslide nowcasting. Sensors. 2020;**20**:1425. DOI: 10.3390/s20051425

[25] Li B, Li R, Sun T, et al. Improving LSTM hydrological modeling with spatiotemporal deep learning and multi-task learning: A case study of three mountainous areas on the Tibetan plateau. Journal of Hydrology. 2023;**620**:129401. DOI: 10.1016/j.jhydrol.2023.129401

[26] Gupta A. Information and disinformation in hydrological data across space: The case of streamflow predictions using machine learning. Journal of Hydrology: Regional Studies. 2024;**51**:101607. DOI: 10.1016/j.ejrh.2023.101607

[27] Hu F, Yang Q, Yang J, et al. Incorporating multiple grid-based data in

CNN-LSTM hybrid model for daily runoff prediction in the source region of the Yellow River Basin. Journal of Hydrology: Regional Studies. 2024;**51**:101652. DOI: 10.1016/j.ejrh.2023.101652

[28] Kratzert F, Klotz D, Shalev G, et al. Towards learning universal, regional, and local hydrological behaviors via machine learning applied to large-sample datasets. Hydrology and Earth System Sciences. 2019;**23**:5089-5110. DOI: 10.5194/hess-23-5089-2019

[29] Kratzert F, Nearing G, Addor N, et al. Caravan - a global community dataset for large-sample hydrology. Scientific Data. 2023;**10**:61. DOI: 10.1038/s41597-023-01975-w

[30] Arnold JG, Srinivasan R, Muttiah RS, et al. Large area hydrologic modeling and assessment part I: Model development. Jawra. Journal of the American Water Resources Association. 1998;**34**:73-89. DOI: 10.1111/j.1752-1688.1998.tb05961.x

[31] Lee J, Hwang S. Ungauged basin flood prediction using long short-term memory and unstructured social media data. Water. 2023;**15**:3818. DOI: 10.3390/w15213818

[32] Xiao Q, Zhou L, Xiang X, et al. Integration of hydrological model and time series model for improving the runoff simulation: A case study on BTOP model in Zhou River basin, China. Applied Sciences. 2022;**12**:6883. DOI: 10.3390/app12146883

[33] Xie K, Liu P, Zhang J, et al. Physics-guided deep learning for rainfall-runoff modeling by considering extreme events and monotonic relationships. Journal of Hydrology. 2021;**603**:127043. DOI: 10.1016/j.jhydrol.2021.127043

[34] Nimai S, Ren Y, Ao T, et al. Enhancing runoff simulation using BTOP-LSTM hybrid model in the Shinano River basin. Water. 2023;**15**:3758. DOI: 10.3390/w15213758

[35] Sun R, Yuan H, Yang Y. Using multiple satellite-gauge merged precipitation products ensemble for hydrologic uncertainty analysis over the Huaihe River basin. Journal of Hydrology. 2018;**566**:406-420. DOI: 10.1016/j.jhydrol.2018.09.024

[36] Jiang P, Shuai P, Sun A, et al. Knowledge-informed deep learning for hydrological model calibration: An application to Coal Creek watershed in Colorado. Hydrology and Earth System Sciences. 2023;**27**:2621-2644. DOI: 10.5194/hess-27-2621-2023

Section 4

Nature-Based Solutions

Chapter 6

Nature-Based Solutions for River Restoration and Flow Management: The Case of Kitwe City, Zambia

Yakob Umer, Sisay E. Debele, Chizya Mvula, Giriraj Amarnath, Moses N. Chisola and Belen Marti-Cardona

Abstract

River systems worldwide are under significant anthropogenic pressures and climate-related challenges, leading to ecosystem degradation and increased flood risk. This chapter demonstrates how Nature Based Solutions (NbS) can contribute to river restoration while reducing flood risk, supporting wider sustainable goals. To this end, this chapter evaluates the effectiveness of NbS interventions in river restoration and flood risk management in the Kitwe City, Zambia. The methodology involves using a hydraulic model to simulate river flow under different NbS scenarios (retention ponds and woodland reforestation), and to compare the simulated flood depth and flow velocity in pre- and post-intervention conditions. The findings indicate that the presence of NbS significantly reduces flood risks, with retention ponds and woodlands leading to flood depth reductions ranging from 0.09 m to 0.18 m and 0.06 m to 0.11 m, respectively. Regarding flow velocities, retention ponds reduced them by an average of 0.11m/s, and woodlands, by 0.07 m/s. These results indicate that both NbS types reduce flood depth and velocity, with ponds being slightly more effective than woodland in the particular setting of the Kitwe District. The findings suggest that integrating NbS into river restoration can mitigate flood risks, improve ecosystem resilience, and contribute to long-term sustainability. These results inform risk management policies and emphasise the need for interdisciplinary collaboration to upscale NbS for maximum ecological and societal benefits.

Keywords: climate change, nature-based solutions, flood risk reduction, flood depth and velocity, nature-based solution policies and regulations

1. Introduction

1.1 Overview of river systems and their importance

Rivers play an indispensable role in supporting both natural ecosystems and human activities. They provide essential ecosystem services, such as freshwater supply, nutrient cycling, habitat provision, and flood regulation [1]. Rivers are crucial for agriculture, industries, and domestic purposes, supporting the livelihoods of millions globally [2]. In addition to their economic importance, rivers are cultural and recreational resources, holding significant social and aesthetic value. However, global river systems have been degraded due to human-induced urbanisation, industrialisation, and deforestation [3].

Despite their importance, rivers are under immense pressure from pollution, over-abstraction, and the impacts of climate change [4]. Agricultural runoff, untreated sewage, and industrial discharges are among the leading causes of water pollution, contributing to water quality degradation and biodiversity loss [5]. Furthermore, the over-abstraction and regulation of using reservoirs to ensure water supply for irrigation, urban consumption, and industrial use have significantly altered the natural flow regimes of many river systems, endangering aquatic ecosystems [6, 7]. Climate change compounds these issues, increasing the frequency and intensity of extreme weather events like floods and droughts, further threatening river systems and the communities that depend on them [8].

1.2 The need for river restoration and flow management

Historically, river systems have been heavily modified through dam construction, channelisation, and levee-building to support irrigation, hydropower generation, and flood control [9]. While these interventions have facilitated economic development, they have also disrupted natural flow patterns, leading to environmental degradation, including reduced biodiversity, habitat loss, and increased vulnerability to extreme climatic events [10]. As the ecological impacts of these modifications have become clearer, there has been a growing shift towards restoring rivers to their natural flow regimes to enhance ecological integrity and ecosystem resilience [11].

Restoration efforts aim to reestablish natural river dynamics, improve water quality, support biodiversity, and enhance resilience to climate change [12]. Effective river restoration and flow management are thus crucial for balancing ecological health with human needs. In this context, Nature-Based Solutions (NbS) have emerged as a promising strategy to achieve sustainable, long-term outcomes [13].

The challenges faced in the Kitwe City, located upstream of the Kafue River, Zambia, exemplify the need for such solutions. Kitwe experiences annual flooding between December and April due to overflow from the Kafue River as a result of heavy rainfall in the headwaters and direct precipitation over the region [14]. Human activities, such as deforestation and urban encroachment, have further altered the flood dynamics, increasing the area's vulnerability. Given these factors, Kitwe provides an ideal case study to test and validate the effectiveness of NbS for flood risk management. Thus, this study addresses the following research question: do NbS support restoration and flow management, and can their effects be evaluated? This study aims to assess the effectiveness of different NbS interventions in reducing flood risk in Kitwe City, Zambia. To achieve this, we developed and calibrated a 2D) hydraulic model of the current conditions. We then incorporated various proposed NbS interventions, such as woodlands and retention ponds, into the model. All models were run using the same boundary conditions (the hydrograph at Smith's Bridge). The effectiveness of the NbS was evaluated by comparing the spatial and temporal changes in water depth and velocity resulting from their implementation.

2. Principles of river restoration and flow management using NbS

NbS are increasingly recognised as key strategies for river restoration and sustainable flow management. These approaches integrate ecological principles and human well-being to enhance river ecosystems while addressing critical challenges, such as flood risks, water scarcity, and biodiversity loss [15]. The following principles underpin the successful implementation of NbS in river restoration and flow management, emphasising ecological integrity, ecosystem function, and socio-ecological integration.

2.1 Defining nature-based solutions

NbS are approaches that utilise natural processes and ecosystem services to address societal challenges, such as water management, biodiversity conservation, and climate change adaptation [16]. In the case of river restoration, NbS focus on restoring natural hydrological processes, such as floodplain reconnection, wetland restoration, and reforestation of riparian zones [7]. These approaches help manage water resources sustainably and enhance ecosystem resilience, contributing to biodiversity conservation [2].

NbS for river restoration and flow management contribute to several United Nations Sustainable Development Goals (SDGs) and align with the European Union's (EU's) Green Deal. Specifically, they address SDG 6 (Clean Water and Sanitation) by improving water quality and ensuring sustainable water management, and SDG 13 (Climate Action) by enhancing resilience against climate-related disasters like floods and droughts. NbS also support SDG 14 (Life Below Water) by restoring freshwater ecosystems and SDG 15 (Life on Land) by promoting biodiversity and preventing land degradation. Additionally, they contribute to SDG 11 (Sustainable Cities and Communities) by making urban areas more resilient to flooding and enhancing green spaces. In alignment with the EU Green Deal, NbS contribute to biodiversity conservation, climate adaptation, and water management objectives [13, 17].

2.2 Ecological integrity and flow dynamics

Restoring the ecological integrity of rivers requires a deep understanding of their natural hydrological cycles. Ecological integrity refers to the preservation of a river's natural processes, including hydrological, geomorphological, and biological functions [18]. A crucial aspect of this integrity is the flow regime, which regulates the movement of water, sediment, and nutrients through the river system.

The natural variability in river flow, both seasonally and annually, plays a vital role in sustaining aquatic and riparian biodiversity. For instance, flood events can help to maintain floodplain ecosystems, create habitat heterogeneity, and trigger reproductive cycles for many aquatic species [19]. Conversely, flow reductions, due to human interventions, such as dams and over-extraction, can fragment habitats and disrupt ecological processes [9]. NbS aim to restore natural flow dynamics through interventions, such as floodplain reconnection, removal of obsolete dams, and reforestation of upstream areas, to enhance water retention and reduce flood peaks [20]. Maintaining natural flow regimes through NbS can also improve ecosystem resilience to climate change. As global warming intensifies, river systems are increasingly subjected to extreme hydrological conditions, such as prolonged droughts and intense floods. NbS help to buffer these impacts by restoring natural hydrological cycles and improving water retention in landscapes [7].

2.3 The role of ecosystem functions in river health

Healthy river ecosystems rely on the functioning of various interconnected habitats, including riparian zones, wetlands, and floodplains. These areas act as natural buffers that mitigate the impacts of floods, filter pollutants, and support biodiversity [16]. Restoring these habitats is critical to maintaining the ecological health of rivers. Riparian zones, for example, stabilise riverbanks, reduce erosion, and provide habitat for both aquatic and terrestrial species. Vegetation in these zones helps to filter nutrients and sediments from surface runoff, improving water quality and supporting aquatic life [21]. Wetlands and floodplains further contribute by absorbing excess floodwaters, reducing downstream flood risks, and recharging groundwater [22].

NbS interventions often focus on the restoration of these key ecosystems. For instance, reforesting riparian areas, re-creating wetlands, or removing embankments that disconnect rivers from their floodplains can significantly enhance the overall health of river systems [23, 24]. These interventions are cost-effective and sustainable compared to traditional "grey" infrastructure solutions, such as levees or concrete channels, which often result in long-term ecological degradation [11]. Upstream interventions can significantly impact downstream conditions. For example, restoring riparian zones through reforestation stabilises riverbanks, reduces erosion, and filters pollutants. This improvement in water quality and habitat conditions can be observed downstream [21].

Similarly, wetland restoration absorbs excess floodwaters, recharges groundwater, and filters contaminants, which mitigates flood risks and enhances water quality downstream [22]. Reconnecting rivers to their natural floodplains by removing levees or embankments allows rivers to spread out during high flows, reducing flood intensity and creating essential habitats [23, 24]. Additionally, removing dams restores natural river flows, improves sediment transport, and facilitates fish migration, benefiting ecosystems both upstream and downstream [11].

2.4 Integrating socio-ecological systems

Rivers provide ecosystem services that benefit human populations, including water supply, recreation, and cultural values [2]. Therefore, successful NbS interventions must consider the needs and perspectives of local communities and stakeholders who rely on river systems. Engaging local communities in river restoration projects ensures that NbS are socially acceptable and economically viable [15]. Local stakeholders often possess traditional ecological knowledge, which can inform sustainable management practices. Moreover, involving communities in restoration efforts helps to foster a sense of stewardship and responsibility towards the ecosystem [25]. Examples of this include participatory river restoration projects, where community members are actively involved in reforestation, wetland creation, or monitoring of water quality. Balancing ecological restoration with economic and social objectives is vital for the long-term success of NbS. For instance, floodplain restoration can reduce flood risks for communities while enhancing biodiversity, but it may also require changes in land-use or agricultural practices. In such cases, compensation schemes or incentives for sustainable farming can help align ecological goals with local livelihoods [18].

3. Nature-based solutions for river restoration and flow management

3.1 Nature-based solutions for river restoration

NbS for river restoration utilise natural processes and ecosystems to revive river systems and enhance their ecological functions. These approaches aim to mitigate environmental degradation, improve water quality, and support biodiversity. Among the key NbS for river restoration are riparian buffer zones, which involve planting vegetation along riverbanks. These buffers stabilise soils, filter pollutants, and provide essential habitat, thus reducing erosion and improving water quality by trapping sediments and nutrients from runoff [26]. Another important approach is wetland restoration, which includes rehabilitating or creating wetlands to enhance flood attenuation, water filtration, and habitat provision. Wetlands function as natural sponges that absorb excess nutrients and mitigate floods [27].

Floodplain reconnection is also a significant NbS, involving the restoration of natural floodplain functions by reconnecting rivers with their floodplains. This strategy allows rivers to spread out during high flows, which help reduce flood risks and improve aquatic habitat [28]. Additionally, streambank stabilisation using natural materials and vegetation addresses erosion issues and enhances habitat. Techniques, such as live stakes and fascines, can effectively control erosion and improve habitat conditions [29]. In-stream habitat enhancement, which includes adding structures like boulders and woody debris, also plays a crucial role. These structures provide vital habitats for fish and other aquatic organisms, supporting biodiversity [30].

3.2 Nature-based solutions for flow management

For flow management, NbS focus on harnessing natural processes to manage river flows effectively, thereby mitigating the impact of floods and droughts while supporting ecological health. Riparian and floodplain restoration are central to this approach. These practices involve reestablishing natural flow patterns and enhancing the floodplain's capacity to absorb and slow water, which mitigates flood risks and improves groundwater recharge [31].

Constructing new wetlands, improving existing ones, or using retention and detention ponds can manage runoff and store excess water, capturing and slowing stormwater to reduce peak flow rates and improve water quality [32]. Pervious surfaces and green infrastructure, such as permeable pavements and green roofs, increase infiltration and reduce surface runoff. These solutions help manage stormwater and lessen the load on conventional drainage systems [33]. Reforestation and afforestation contribute by planting trees in catchment areas to boost water infiltration and reduce runoff. Forested areas improve soil structure and water retention, which can help mitigate both floods and droughts [34]. Lastly, natural flood management utilises features like wetlands, floodplains, and forests to manage flood risks by working with natural processes to store and slow floodwaters, thereby reducing impacts on downstream areas [35].

4. Case studies: NbS for river restoration and flow management

4.1 Study area

The city of Kitwe is located in the central part of Zambia, in the Copperbelt Province (**Figure 1**). It is the second largest city in Zambia in terms of population and economic activities. The city has an area of 1205 km^2 at an elevation of 1295 m above sea level. It hosts a population of 661,901 [36].

The Kafue River, which is the biggest tributary of the Zambezi, flows through Kitwe from north to south. Near the banks of the river are several densely populated urban settlements, namely, Luangwa, Ipusukilo, Musonda, and Chipata townships, as shown in **Figure 2**. These settlements are located within the Wusakile Ward, which has a population of 16,468 people residing in 2,745 households, and is an area highly vulnerable to flooding [14]. In January 2023, the area was devastated by severe flooding, triggered by two flood waves with estimated return periods of 5 and 20 years, respectively.

As a response to the flooding caused by the bursting of the banks of the Kafue River in Kitwe, the GloFAS (Global Flood Awareness System) and the Impact-based forecast (IBF) system for floods have primarily been used to predict floods. Locally, the water levels at the upstream gauging station, Kafue at Smith's Bridge, have been used to verify and validate the warning from the system. Depending on the size of the predicted flood and the lead time for the flood, early action protocols may be activated. In recent times, evacuation and relief aid from humanitarian and disaster relief agencies have been executed. However, there is a need for more sustainable and permanent solutions to flooding such as environmental (rainwater harvesting and channelling water away from flooded areas) and nature-based solutions such as afforestation and reforestation.

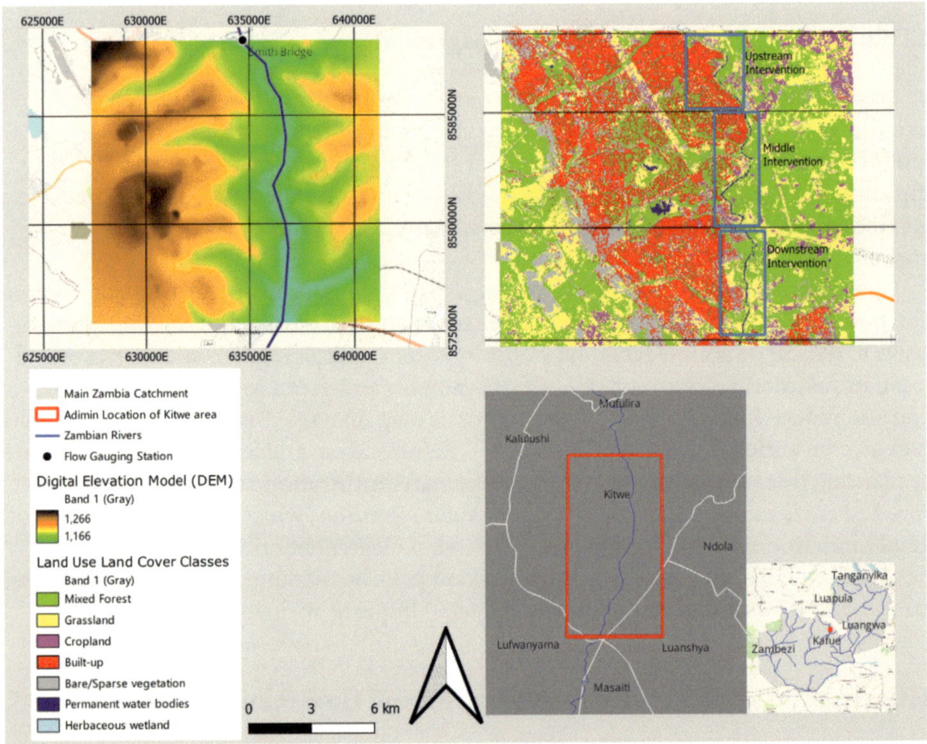

Figure 1.
Location of the case study area and land cover map used in the model. Blue rectangles indicate the location of NbS interventions, where pre-NbS land-use classes changed to different scenarios. The red rectangle indicates the administrative location of the case study in Zambia.

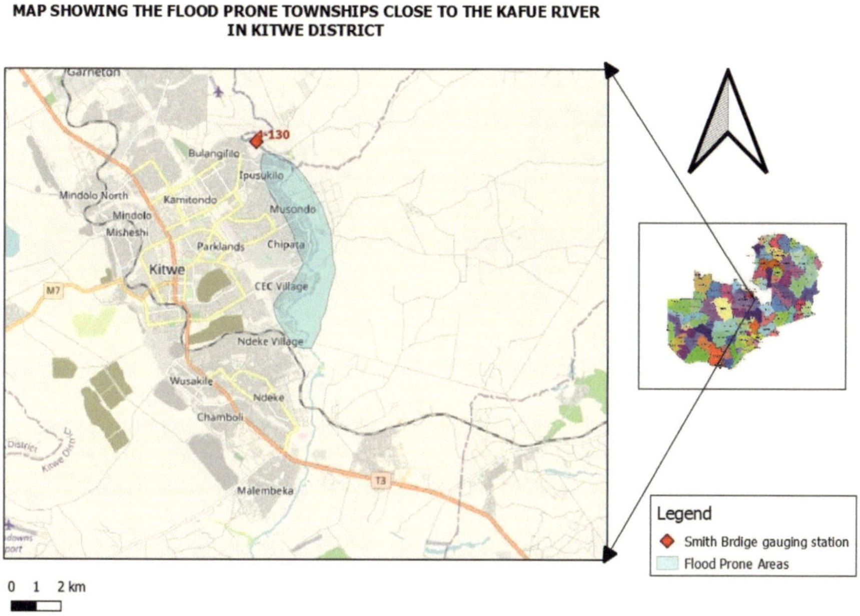

Figure 2.
Location of Kitwe City in Kitwe City that are prone to floods (post-flood information map).

4.2 Data

In this study, we used daily river flow data from the 2023 flood event, as the upstream boundary condition for the HEC-RAS model. The data consisted of water stage measurements collected by the Water Resources Management (WARMA) at the Smith Bridge gauging station. These measurements were converted into flow discharge using the station's rating curve. As illustrated in **Figure 3**, the flood hydrograph starts building up from December 2022 onwards and attains its peak between January and March 2023, with a flood threshold of 200 m^3/s corresponding to a 5-year return period flood event. According to information from WARMA, the 5-year event is considered as the threshold for the flood danger level.

4.3 Methods

The Hydraulic Engineering Center's River Analysis System (HEC-RAS), version 6.5 [37], was set up as a coupled 1D-2D model, where river flow was modelled in one dimension, and floodplain flow was modelled in two dimensions. The digital terrain model (DTM) represents the elevation of the ground surface. HEC-RAS supports DTMs in raster format, where the cell size used must be fine enough to capture the terrain's features. The terrain data are used in HEC-RAS to visualise the floodplain geometry and calculate the water movement on it. To account for changes in land cover within the HEC-RAS model (a 2D unsteady hydraulic model), we used high-resolution (10 m) land cover data derived from Sentinel-2 satellite optical imagery. Manning's n values were assigned to the land-use categories shown in **Figure 1**. Roughness coefficients (Manning's n), representing the surface's resistance to flow, are critical parameters for calculating water depth,

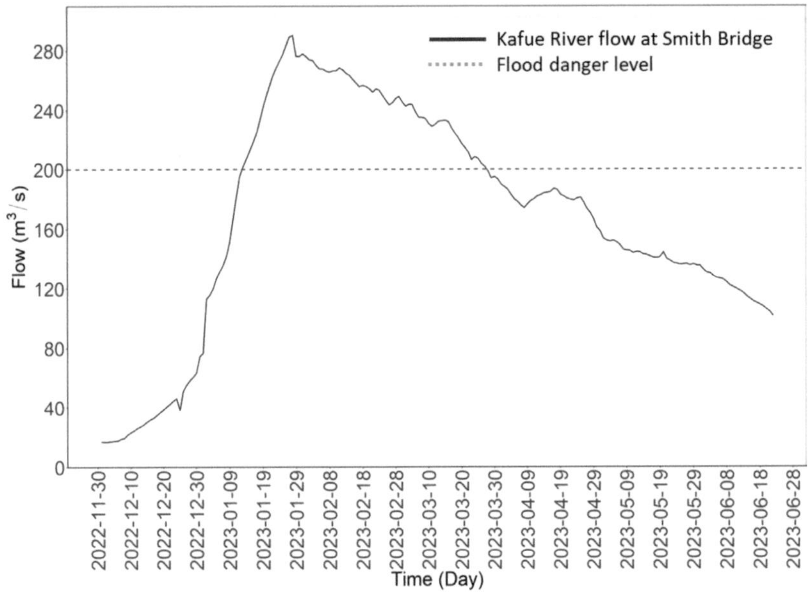

Figure 3.
Flow hydrograph for the flood year 2023. The flow was used as the upstream boundary condition for the HEC-RAS model.

surface elevation, and velocity under both pre-NbS (base model) and post-NbS conditions.

Different scenarios were modelled: pre-NbS, representing conditions without any NbS in place, and post-NbS, representing two different NbS strategies—water-based (ponds) and vegetation-based (forest). These strategies were applied to the upstream, midstream, and downstream sections (**Figure 1**). We conducted a post-NbS simulation by updating the 2D base model, keeping all other parameters consistent with the initial base simulation. The results of this simulation are shown in **Figures 4** and **5**. To assess the benefits of the NbS, we compared the two simulations by subtracting the post-NbS scenario from the pre-NbS scenario (base model). The differences between these two scenarios are analysed and presented in **Figure 4**. The impact of the NbS is evident in the reduced flood velocity and improved floodwater retention within the main channel, as reflected by the increased water depth in the NbS simulation compared to the base model (**Figure 5**).

4.4 Results and discussion

The results in **Figure 4a–d** show spatial changes in water depth and velocity calculated as a difference between the simulation without NbS and with NbS of woodlands and retention ponds. In **Figure 4a,b**, the intervention scenarios lead to a reduction in water depth (negative depth), particularly in the upstream regions and the periphery of the downstream sections of the river. In the middle and lower reaches of the river, however, the water depth shows a slight increase. Both scenarios display similar spatial patterns along the river. Woodlands contribute to a more pronounced increase in water depth (0.8 to 1.2 m) in the downstream sections due to the gradual accumulation of water over time. In contrast, retention ponds cause a greater reduction in water

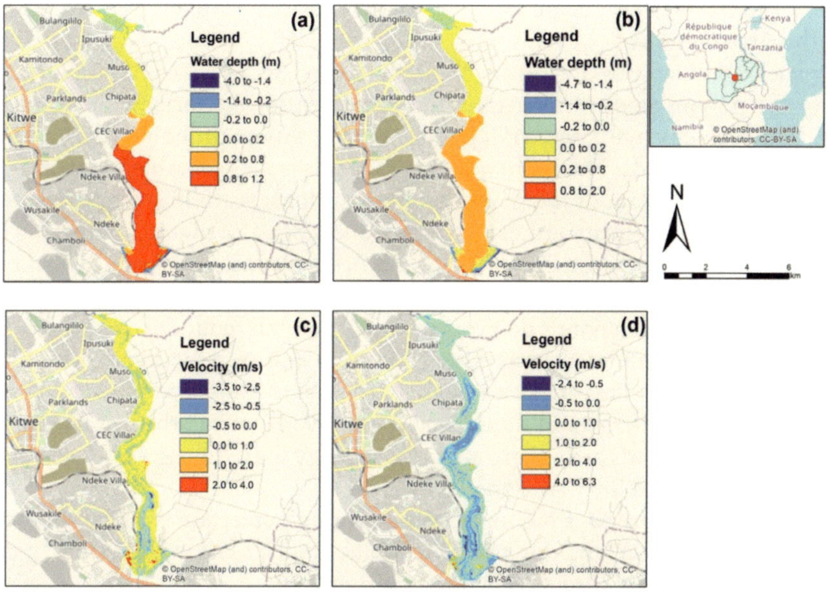

Figure 4.
Water depth and velocity differences: (a) water depth differences between pre-NbS and woodland simulations, (b) water depth differences between pre-NbS and ponds simulations; (c) velocity differences between pre-NbS and woodland simulations, and (d) velocity differences between pre-NbS and ponds simulations.

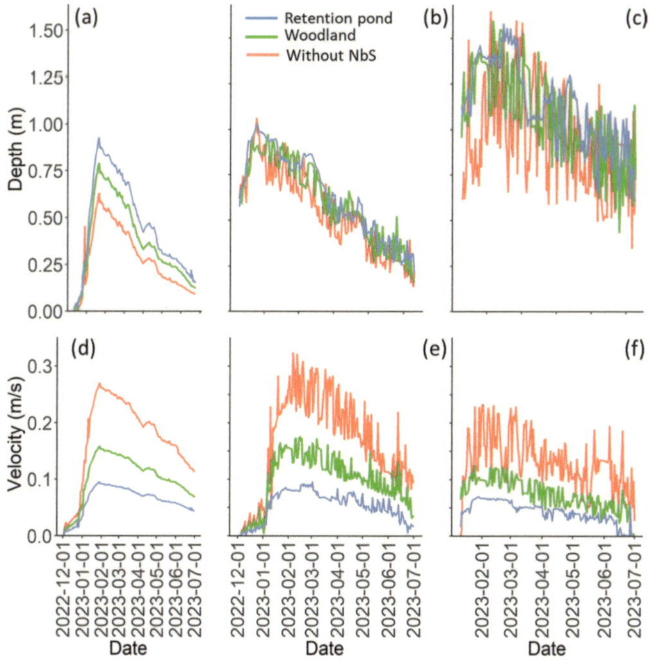

Figure 5.
Time series of depth (m) and velocity (m/s) at upstream (left panel), midstream (middle panel), and downstream (right panel) for three scenarios: Retention pond (blue), woodland (green), and without NbS (red). Panels (a)–(c) represent depth measurements at upstream, midstream, and downstream locations, while panels (d)–(f) depict velocity data at the same respective locations.

depth (0.2 to 0.8 m) in the same areas, primarily due to the attenuation of floodwaters along the river sections.

Figure 4c,d illustrates the impact of Nature-Based Solutions (NbS) on flood velocity along inundated river sections. Both intervention scenarios significantly reduce flood velocity, with retention ponds exerting a more pronounced effect compared to woodlands. The influence of woodlands is most notable in the middle and downstream sections of the river. However, for most sections, the changes are minimal, with flood velocity either remaining unchanged or slightly increasing (0 to 1 m/s), suggesting that additional roughness is needed to achieve a substantial reduction in velocity.

In contrast, retention ponds demonstrate a substantial reduction in flood velocity, especially in larger inundated areas downstream, as well as in sections of the middle and upstream regions. This effect is largely attributed to their capacity to store water, which decreases the energy and flow speed of the river. Overall, the results suggest that retention ponds are a more effective strategy for reducing flood velocities compared to woodland interventions.

The results in **Figure 5a–f** highlight how retention ponds and woodlands, located at upstream, midstream, and downstream points along a river, effectively reduce flood depth and velocity compared to areas without NbS.

In **Figure 5(a–c)**, the depth of the river is consistently lower in scenarios with retention ponds (blue line) and woodlands (green line) compared to areas without NbS (red line). This pattern is observed across all locations along the river (upstream, midstream, and downstream). The retention ponds show the greatest reduction in flood depth, especially at the upstream and midstream sections, by temporarily holding back excess river water during peak flood events and releasing it slowly over time. Woodlands also provide substantial flood depth reduction, particularly in midstream and downstream areas, by enhancing soil infiltration, increasing water retention in the landscape, and slowing down runoff. In contrast, areas without NbS suggest that the absence of flood mitigation strategies leads to faster and higher floodwaters (**Figure 4a–c**).

Figure 5(d–f) shows similar patterns for flow velocity. Retention ponds and woodlands both reduce the speed at which water moves through the river system, with the retention ponds (blue) significantly decreasing velocities at all points. This is particularly evident in the upstream and midstream sections, where ponds act as temporary storage areas, reducing the energy and speed of floodwaters. Woodlands (green line) reduce velocity to a lesser extent but still show a marked difference compared to areas without NbS (red line), especially in downstream regions. The vegetation in woodlands acts as a natural barrier that increases roughness and intercepts rainfall, slowing down the water flow. In contrast, the areas without NbS experience consistently higher velocities across all sections of the river, which increase the risk of erosion, channel instability, and downstream flooding.

Figure 5 also demonstrates the spatial effectiveness of NbS, indicating that the impact of these interventions varies along different sections of the river. Retention ponds have the greatest effect in upstream and midstream areas, where they help prevent rapid downstream flooding by capturing and gradually releasing water. Woodlands are particularly effective in midstream and downstream sections, where their capacity to disperse water and increase infiltration helps prevent the accumulation of floodwaters. In contrast, without NbS, the river system experiences higher and faster flood peaks, especially in downstream sections, where the cumulative impact of upstream runoff is most pronounced.

On average, retention ponds and woodlands, placed strategically along upstream, midstream, and downstream sections of the river, are highly effective at reducing flood depth and velocity. Overall, the flood depth reaches 0.60 m without NbS and 0.72 m in areas equipped with NbS, resulting in an overall reduction of 0.11 m in flood depth. Similarly, the flood velocity reaches 0.16 m/s without NbS and decreases to 0.07 m/s with NbS in place, with an overall reduction of 0.09 m/s. The flood depth is reduced by 0.09 m in woodland areas and 0.18 m in retention ponds upstream, 0.06 m (woodland) and 0.09 m (retention ponds) midstream, and 0.11 m (woodland) and 0.17 m (retention ponds) downstream, compared to areas without NbS (**Figure 5a–c**). This reduction is due to the retention of water within the main channel, preventing it from quickly flowing out and allowing for increased infiltration and a rise in the groundwater table. Similarly, flood velocity is reduced by 0.07 m/s (woodland) and 0.11 m/s (retention ponds) upstream, 0.07 m/s (woodland) and 0.11 m/s (retention ponds) midstream, and 0.06 m/s (woodland) and 0.09 m/s (retention ponds) downstream, compared to areas without NbS (**Figure 5d–f**). This reduction in velocity results from slowing down the floodwater, preventing it from flowing rapidly onto the floodplain along the main river.

These findings align with research that emphasises the role of upstream and midstream NbS in managing downstream flood risks. Strategically placed retention ponds in river catchments have been shown to reduce flood peaks by up to 50%, as they slow and regulate the timing of water flow [38]. Similarly, woodlands help reduce both water depth and velocity by increasing natural infiltration and introducing roughness that disrupts water flow [39]. This approach of positioning NbS upstream and midstream supports natural water management strategies that aim to control river flooding at its source and prevent large, fast-moving flood waves from impacting downstream communities.

This study reveals that retention ponds significantly lower water velocities at all points along the river, especially in the upstream and midstream areas. This observation aligns with the research by Hamel et al. [38], which indicates that strategically placed retention ponds can decrease flood peaks by up to 50%. Such ponds serve as temporary storage areas, effectively regulating water flow and timing, thereby playing a crucial role in managing flood dynamics. In contrast, while woodlands also contribute to reduced water velocities, their impact is less pronounced than that of retention ponds. This is supported by O'Donnell et al. [39], who emphasise that the roughness and increased infiltration offered by woodlands slow down water flow, albeit to a lesser extent than engineered solutions. The effectiveness of these interventions varies spatially, as demonstrated in the study. Retention ponds exert the most significant influence in upstream and midstream areas, while woodlands show greater efficacy in midstream and downstream regions. This spatial variability is a common theme in NbS research. For instance, Fletcher et al. [40] highlight the importance of site-specific strategies in enhancing water management through the placement of green infrastructure that accounts for local hydrological conditions.

Moreover, the study reports a clear reduction in flood depths in areas equipped with NbS compared to those without. This finding aligns with the work of Pettit et al. [41], which demonstrates that the integration of natural features into river systems can lead to substantial decreases in flood depth and associated risks. Their research underscores how various NbS, including wetlands and vegetation, contribute to increased water retention and slower runoff, further reinforcing the value of these approaches. Moreover, the study reports a clear increase in flood depth in areas equipped with NbS compared to those without NbS. This highlights the importance of using multiple NbS in combination, such as wetlands and floodplain restoration, to reduce both flood velocity and depth simultaneously [42, 43].

NbS serve as retention of water within the main channel, preventing it from quickly flowing out and allowing for increased infiltration and a rise in the groundwater table. This effect is supported by findings from Barton et al. [44], who argue that natural systems can enhance groundwater recharge significantly, thereby mitigating flood risks and improving water quality. This dual benefit—reducing flood risk while replenishing groundwater—strengthens the argument for employing NbS in integrated water resource management. Furthermore, the approach of utilising NbS as a sustainable method for flood management is gaining recognition in contemporary research. A recent meta-analysis by Mazzorana et al. [45] illustrates that NbS not only provide immediate flood control benefits but also foster long-term ecological resilience and community adaptation to changing climatic conditions. This notion is reflected in the findings from **Figure 5**, which support the integration of NbS into flood management strategies, leading to effective and sustainable solutions.

5. Challenges and limitations of implementing and upscaling of NbS

The successful implementation and scaling of NbS in the Kitwe region of Zambia face several challenges and limitations. These include resource constraints, policy gaps, lack of coordination among stakeholders, and community engagement issues.

One of the most pressing challenges is limited financial and technical resources. NbS interventions, such as wetland restoration, reforestation, and retention pond construction, require significant investment not only for their development but also for ongoing maintenance and monitoring. In regions like Kitwe, where economic resources are already stretched thin, priority is often given to traditional "grey infrastructure" solutions, such as levees and dams, which receive more political and financial backing, despite their long-term ecological costs [46]. Additionally, there is a shortage of local expertise and technical knowledge necessary for the design and implementation of NbS interventions, further impeding their widespread adoption [26].

5.1 Policy and regulatory limitations

Current water management policies in Zambia often do not fully recognise or prioritise NbS approaches within flood risk management frameworks. This gap in governance limits the potential for NbS to be integrated into national or local plans [39]. Policies tend to focus on engineering-based solutions, which are easier to justify politically and financially but often fail to provide the same long-term ecological and societal benefits as NbS. In addition, the lack of incentives or supportive regulations for adopting NbS prevents stakeholders from fully embracing these sustainable solutions [11].

5.2 The lack of collaboration and coordination among stakeholders

Effective NbS interventions require the involvement of various groups, including government agencies, local communities, non-governmental organisations (NGOs), and the private sector. However, in the Kitwe region, these stakeholders often operate in silos, leading to fragmented efforts and conflicting objectives. This lack of coordination can result in poorly integrated projects that fail to deliver the expected benefits [15]. Additionally, conflicting land-use priorities, such as agriculture versus environmental conservation, often hinder the smooth implementation of NbS [25].

5.3 Community engagement and acceptance of NbS

Local communities, whose livelihoods often depend on agriculture or other land uses, may be resistant to changes in land management practices that NbS might necessitate, such as reforestation or the restoration of floodplains. Without proper education and engagement, communities may not understand the long-term benefits of NbS or may view these interventions as a threat to their immediate economic needs. Ensuring community involvement and providing incentives or compensation for changes in land use are critical for overcoming these barriers [18, 47].

5.4 Climate uncertainty and variability pose additional risks

While NbS can increase resilience to climate change by enhancing water retention and reducing flood peaks, the unpredictability of future climatic conditions complicates the design of these solutions. For instance, changing rainfall patterns and the increasing frequency of extreme weather events may affect the effectiveness of certain NbS interventions, such as retention ponds or woodlands, which depend on stable hydrological conditions [8].

5.5 Lack of site, region, and climate-specific monitoring and evaluation methods

Robust monitoring systems are essential for tracking the long-term effectiveness of these interventions and ensuring they deliver the anticipated benefits. However, in Kitwe, the limited capacity for ongoing data collection and analysis hampers the ability to assess the success of NbS projects. Without proper monitoring, it is difficult to make informed adjustments to NbS strategies or demonstrate their value to policymakers and funders [26]. In summary, while NbS offer promising solutions for river restoration and flood management in the Kitwe region, these challenges must be addressed to ensure their successful implementation and upscaling. A concerted effort involving strengthened policies, increased funding, better stakeholder coordination, and robust community engagement is essential to overcome these barriers and fully realise the potential of NbS under changing climate conditions.

6. Future directions and recommendations

The successful implementation of NbS for river restoration and flood management requires a forward-thinking approach that integrates ecological, social, and economic aspects. To maximise the potential of NbS, a combination of general and specific recommendations is drawn as follows:

- A critical step in the future development of NbS is engagement with stakeholders. This involves fostering close collaboration between local communities, policymakers, non-governmental organisations (NGOs), and the private sector. A co-creation model, where stakeholders are involved in both identifying problems and implementing solutions, will enhance the acceptance and effectiveness of NbS. Local communities, in particular, should play a central role, as they possess valuable ecological knowledge and are directly affected by river management practices. Workshops and training programmes should be established to raise awareness of the benefits of NbS over conventional "grey" infrastructure, building technical capacity and fostering local ownership of these solutions.

- Integrating NbS into national and local water management policies is paramount. Current regulatory frameworks often emphasise engineering-based solutions, which may offer short-term relief but fail to provide the long-term sustainability benefits that NbS can deliver. Advocacy efforts should focus on embedding NbS into policies, aligning them with broader SDGs, particularly SDG 6 (Clean Water and Sanitation) and SDG 13 (Climate Action).

- Scaling up NbS interventions also requires a methodical approach. Pilot testing in diverse environments should be pursued to evaluate the effectiveness of different strategies. These pilot projects should be paired with robust monitoring and evaluation systems to track their impacts over time. Continuous data collection will allow for adaptive management, ensuring that interventions remain effective as environmental conditions change. Increased funding will be essential for scaling up NbS, as significant investment is required for both development and ongoing maintenance. Financial mechanisms, such as incentives for landowners and farmers to engage in sustainable land-use practices like reforestation and wetland restoration, should be explored.

- Multidisciplinary approaches must be prioritised. NbS for river restoration and flow management inherently require collaboration between ecologists, hydrologists, engineers, and social scientists. This interdisciplinary framework ensures that solutions not only address ecological health but also consider social and economic needs, particularly in communities that rely on rivers for their livelihoods.

- Future work needs to consider the application of data-driven decision-making. Comprehensive flood risk analyses should be conducted to assess the socio-ecological impacts of flooding under various scenarios. Using hydraulic models and climate projections, decision-makers can evaluate the effectiveness of NbS in mitigating flood risks under different climate and land-use change conditions. By incorporating future climate scenarios into planning, NbS can be designed to provide resilience against increasing extreme weather events, such as prolonged droughts and more frequent floods. This will ensure that NbS interventions are robust and adaptable to the challenges posed by climate change.

- Integrating NbS with other flood management strategies can enhance overall resilience. A hybrid approach that combines NbS with traditional "grey" infrastructure, such as levees or dams, may be necessary in some cases to optimise flood protection while maintaining ecological integrity. Additionally, flood early warning systems and other proactive measures can complement NbS, providing communities with more tools to manage and mitigate flood risks.

- Upscaling these solutions in regions like Kitwe, Zambia, faces several challenges. Limited financial and technical resources, policy gaps, and a lack of coordination between stakeholders often hinder their successful implementation. Addressing these barriers will require concerted efforts to improve policy frameworks, secure dedicated funding, and engage local communities in meaningful ways. Community education, incentives for sustainable land management, and targeted training can help overcome resistance to change and foster greater acceptance of NbS interventions.

7. Summary and conclusions

This chapter examined the effectiveness of NbS for river restoration and flood management, particularly focusing on their application in Kitwe, Zambia. The key objectives were to evaluate how NbS interventions, such as retention ponds and woodland reforestation, can mitigate flood risks while enhancing ecological resilience. HEC-RAS was used to simulate river flow under different NbS scenarios, providing insights into changes in flood depth, velocity, and overall system resilience.

The study found that NbS, particularly retention ponds and woodland reforestation, significantly reduced flood risks. Retention ponds were more effective in reducing both flood depth and velocity, especially in upstream and midstream sections, while woodland interventions proved to be particularly valuable in downstream areas. The results demonstrated that NbS not only mitigated flood risks but also contributed to broader sustainable development goals, including water quality improvement and biodiversity conservation. The key conclusions are drawn as follows:

- Rivers are essential to both ecosystems and human activities but are under pressure from human-induced changes and climate challenges. NbS offer a sustainable approach to restoring these systems while balancing ecological integrity and human needs. NbS, such as reforestation, floodplain reconnection, and wetland restoration, enhance ecological functions, support biodiversity, and provide long-term resilience to climate change.

- In the Kitwe region, retention ponds reduced flood depth by 0.18 m upstream, 0.09 m midstream, and 0.17 m downstream, demonstrating their effectiveness in flood risk mitigation. Woodlands, while less effective than retention ponds in some areas, provided important benefits in downstream flood reduction by slowing runoff and enhancing soil infiltration.

- Retention ponds showed the greatest reduction in flood velocity, decreasing it by 0.11 m/s upstream, 0.11 m/s midstream, and 0.09 m/s downstream. Woodland interventions also reduced velocity, but to a lesser extent, with reductions of 0.07 m/s upstream, 0.07 m/s midstream, and 0.06 m/s downstream. Overall, the reduction in flood velocity highlights the ability of NbS to slow water flow, preventing rapid downstream flooding and erosion.

- The spatial effectiveness of Nature-Based Solutions (NbS) varies significantly across different sections of the river system. Retention ponds were most effective in reducing flood depth and velocity in upstream and midstream areas by temporarily holding excess water and releasing it gradually, preventing rapid downstream flooding. Woodlands, on the other hand, were particularly effective in the downstream sections, where their ability to increase soil infiltration and introduce roughness helped slow runoff and reduce flood peaks. Therefore, it is advised to optimise the allocation of NbS to ensure their maximum benefits and effectiveness, considering both spatial and temporal scales.

While this study highlights the value of NbS for flood management, several challenges persist, including resource constraints, policy gaps, and the need for enhanced stakeholder coordination. Additionally, the unpredictability of climate change introduces uncertainty in the effectiveness of NbS interventions over time. To address these challenges, several recommendations for future work are proposed. First, policy integration is crucial; greater efforts are needed to incorporate NbS into national and local water management policies to ensure their long-term sustainability. Second, stakeholder collaboration is essential. Engaging local communities, governments, and NGOs in the co-creation of solutions will enhance both the acceptance and effectiveness of NbS projects. Third, monitoring and evaluation of NbS should be prioritised. Implementing long-term monitoring systems is necessary to assess the effectiveness of NbS interventions and enable adaptive management. Finally, testing hybrid solutions should be a focus of future research. Exploring combinations of NbS with traditional infrastructure can help maximise flood protection under varying climate conditions.

Acknowledgements

This work was carried out with support from the CGIAR Initiative on Diversification in East and Southern Africa, Ukama-Ustawi. We would like to thank all funders who supported this research through their contributions to the CGIAR Trust Fund (www.cgiar.org/funders). The CGIAR Initiative on Diversification in East and Southern Africa also known as Ukama Ustawi aims to help smallholders transition to sustainably intensified, diversified, and de-risked agri-food systems based on maize in 12 East and Southern African countries, details are given here: https://www.cgiar.org/initiative/diversificationin-esa/. We also gratefully acknowledge the support of the CGIAR Initiative on Climate Resilience, also known as ClimBeR, and Accelerating Impacts of CGIAR Climate Research for Africa, also known as AICCRA, whose resources and assistance were crucial to the development of this chapter. Sisay E. Debele is funded by Economic and Social Research Council grants: Building resilience to floods and heat in the maternal and child health system in Brazil and Zambia (REACH), Grant Number: ES/Y00258X/1. We also appreciate the insights and collaboration of the Project Team/Collaborators, as well as the encouragement from the Zambia Water Resources Management Authority (WARMA) and the International Water Management Institute (IWMI). The authors acknowledge the use of ChatGPT (AI tool) for language polishing of the manuscript.

Author details

Yakob Umer[1], Sisay E. Debele[2]*, Chizya Mvula[3], Giriraj Amarnath[4], Moses N. Chisola[5] and Belen Marti-Cardona[6]

1 International Water Management Institute, Pretoria, South Africa

2 Department of Global Health and Development, Faculty of Public Health and Policy, London School of Hygiene & Tropical Medicine, London, United Kingdom

3 Water Resources Management Authority (WARMA), Zambia

4 International Water Management Institute, Colombo, Sri Lanka

5 School of Natural Sciences, Department of Geography and Environmental Studies, University of Zambia, Lusaka, Zambia

6 Faculty of Engineering and Physical Sciences, School of Sustainability, Civil and Environmental Engineering, University of Surrey, Guildford, United Kingdom

*Address all correspondence to: sisay.Debele@lshtm.ac.uk; debele.eshetu@gmail.com

References

[1] Vörösmarty CJ, McIntyre PB, Gessner MO, et al. Global threats to human water security and river biodiversity. Nature. 2010;**467**(7315):555-561

[2] Postel SL, Richter BD. Rivers for Life: Managing Water for People and Nature. Washington, D.C., USA: Island Press; 2003

[3] Foley JA, DeFries R, Asner GP, et al. Global consequences of land use. Science. 2005;**309**(5734):570-574

[4] Hoekstra AY, Mekonnen MM. The water footprint of humanity. Proceedings of the National Academy of Sciences. 2012;**109**(9):3232-3237

[5] Scholes RJ, Biggs R. A biodiversity intactness index. Nature. 2005;**434**(7029):45-49

[6] Martí-Cardona B, Prats J, Niclòs R. Enhancing the retrieval of stream surface temperature from Landsat data. Remote Sensing of Environment. 2019;**224**:182-191. DOI: 10.1016/j.rse.2019.02.007

[7] Palmer MA, Reidy CA, Nilsson C, et al. Climate change and the world's river basins: Anticipating management options. Frontiers in Ecology and the Environment. 2008;**6**(2):81-89

[8] IPCC (Intergovernmental Panel on Climate Change). Global Warming of 1.5°C: An IPCC Special Report on the Impacts of Global Warming. Geneva, Switzerland: IPCC (Intergovernmental Panel on Climate Change); 2018

[9] Nilsson C, Reidy CA, Dynesius M, Revenga C. Fragmentation and flow regulation of the world's large river systems. Science. 2005;**308**(5720):405-408

[10] Finer M, Jenkins CN. Proliferation of hydroelectric dams in the Andean Amazon and implications for Andes-Amazon connectivity. PLoS One. 2012;7(4):e35126

[11] WWF. Living Planet Report 2020: Bending the Curve of Biodiversity Loss. Gland, Switzerland World Wildlife Fund; 2020

[12] Wohl E. Floodplains and wood: Biogeomorphic agents in river corridors. Geomorphology. 2013;**200**:98-112

[13] European Commission. Towards an EU Research and Innovation Policy Agenda for Nature-Based Solutions & Re-Naturing Cities. Final Report of the Horizon. Brussels, Belgium: European Commission; 2015a

[14] Zambia Red Cross Society (ZRCS). Lessons Learnt Workshop. Lusaka, Zambia: ZRCS; 2023

[15] Cohen-Shacham E, Walters G, Janzen C, Maginnis S, editors. Nature-Based Solutions to Address Global Societal Challenges. Gland, Switzerland: IUCN International Union for Conservation of Nature; 2016

[16] Tockner K, Stanford JA. Riverine flood plains: Present state and future trends. Environmental Conservation. 2002;**29**(3):308-330

[17] Debele SE, Leo LS, Kumar P, Sahani J, Ommer J, Bucchignani E, et al. Nature-based solutions can help reduce the impact of natural hazards: A global analysis of NBS case studies. Science of the Total Environment. 2023;**902**:165824

[18] Palmer MA, Bernhardt ES, Allan JD, Lake PS, Alexander G, Brooks S, et al.

Standards for ecologically successful river restoration. Journal of Applied Ecology. 2005;**42**(2):208-217

[19] Poff NL, Allan JD, Bain MB, Karr JR, Prestegaard KL, Richter BD, et al. The natural flow regime. Bioscience. 1997;**47**(11):769-784

[20] Rohde S, Schütz M, Kienast F, Englmaier P. River widening: An approach to restoring riparian habitats and plant species. River Research and Applications. 2005;**21**(10):1075-1094

[21] Naiman RJ, Décamps H. The ecology of interfaces: Riparian zones. Annual Review of Ecology and Systematics. 1997;**28**:621-658

[22] Mitsch WJ, Gosselink JG. Wetlands. Hoboken, New Jersey, USA: John Wiley & Sons; 2000

[23] Fischer J, Bennett AF. Riparian zones as habitat corridors for wildlife. Conservation Biology. 2011a;**25**(2):253-261

[24] Fischer J, Bennett G. Nature-Based Solutions for Resilient Ecosystems and Societies. Cambridge, United Kingdom: Cambridge University Press; 2011b

[25] Berkes F, Colding J, Folke C. Rediscovery of traditional ecological knowledge as adaptive management. Ecological Applications. 2000;**10**(5):1251-1262

[26] Palmer MA, Bernhardt ES, Schlesinger P, et al. Ecological restoration of river ecosystems: A global review of concepts and practice. Ecological Applications. 2014;**24**(8):1947-1968

[27] Mitsch WJ, Gosselink JG. Wetlands. Wiley; 2015

[28] Beechie T, Pess G, Roni P. Restoration of ecosystem processes and functions. In: Restoration of Aquatic Ecosystems: Science, Technology, and Public Policy. Hoboken, New Jersey, USA: Wiley; 2013

[29] Gray DH, Sotir RB. Biotechnical and Soil Bioengineering Slope Stabilization: A Practical Guide for Erosion Control. Hoboken, New Jersey, USA: Wiley; 1996

[30] Fausch KD, Torgersen CE, Baxter CV, Gowan C. Landscapes to riverscapes: Bridging the gap between research and conservation of stream fishes. Bioscience. 2002;**52**(6):483-498

[31] Zedler JB, Kercher S. Wetland resources and restoration. Annual Review of Environment and Resources. 2005;**30**:39-74

[32] Davis AP, Hunt WF. Stormwater management for residential and commercial areas. Water Environment Research. 2010;**82**(2):123-135

[33] Berndtsson JC. Green roof performance towards management of runoff. Building and Environment. 2010;**45**(3):825-836

[34] Ellison D, Hines J, Brown M. Trees, forests, and water availability. Nature Sustainability. 2017;**1**:397-405

[35] Pitt M, Anderson C, Haines C. Natural flood management: Challenges and opportunities. Water Management. 2017;**170**(7):1189-1201

[36] Zambia Statistical Agency (Zamsats). Zambia Population Census. Lusaka, Zambia: Zamsats; 2022

[37] Army Corps of Engineers. HEC-RAS: River Analysis System. Version 6.5. Davis, California, USA: Hydrologic Engineering Center; 2024

[38] Ahmadisharaf E, Alamdari N, Tajrishy M, Ghanbari S. Effectiveness

of retention ponds for sustainable urban flood mitigation across range of storm depths in northern Tehran, Iran. Journal of Sustainable Water in the Built Environment. 2021;7(2):05021003. DOI: 10.1061/JSWBAY.0000946

[39] Křeček J, Nováková J, Palán L, Pažourková E, Stuchlík E. Role of forests in headwater control with changing environment and society. International Soil and Water Conservation Research. 2021;**9**(1):143-157. DOI: 10.1016/j.iswcr.2020.11.002

[40] Fletcher TD, Brett M, Deletic A. The role of green infrastructure in sustainable water management: A review of current literature. Water Science & Technology. 2013;**67**(1):1-11. DOI: 10.2166/wst.2012.420

[41] Wingfield T, Macdonald N, Peters K, Spees J, Potter K. Natural flood management: Beyond the evidence debate. Area. 2019;**51**(4):743-751. DOI: 10.1111/area.12535

[42] Turkelboom F, Demeyer R, Vranken L, De Becker P, Raymaekers F, De Smet L. How does a nature-based solution for flood control compare to a technical solution? Case study evidence from Belgium. Ambio. 2021;**50**(8):1431-1445. DOI: 10.1007/s13280-021-01548-4

[43] Wu J, Yang R, Song J. Effectiveness of low-impact development for urban inundation risk mitigation under different scenarios: A case study in Shenzhen, China. Natural Hazards and Earth System Sciences. 2018;**18**(9):2525-2536. DOI: 10.5194/nhess-18-2525-2018

[44] Zhang K, Chui TFM. A review on implementing infiltration-based green infrastructure in shallow groundwater environments: Challenges, approaches, and progress. Journal of Hydrology. 2019;**579**:124089. DOI: 10.1016/j.jhydrol.2019.124089

[45] Keech D, Clarke L, Short C. Nature-based solutions in flood risk management: Unlocking spatial, functional and policy perceptions amongst practitioners in South-West England. Nature-Based Solutions. 2023;**4**:100096. DOI: 10.1016/j.nbsj.2023.100096

[46] Bell CD, Wolfand JM, Panos CL, Bhaskar AS, Gilliom RL, Hogue TS, et al. Stormwater control impacts on runoff volume and peak flow: A meta-analysis of watershed modelling studies. Hydrological Processes. 2020;**34**(14):3134-3152. DOI: 10.1002/hyp.13784

[47] O'Donnell EC, Lamond JE, Thorne CR. Recognising barriers to implementation of blue-green infrastructure: A Newcastle case study. Urban Water Journal. 2017;**14**(9):964-971. DOI: 10.1080/1573062X.2017.1279190